INTERNATIONAL CENTRE FOR MECHANICAL SCIENCES

COURSES AND LECTURES - No. 59

VLATKO BRČIĆ
UNIVERSITY OF BELGRADE

PHOTOELASTICITY IN THEORY AND PRACTICE

WITH APPENDICES BY
AUGUSTO AJOVALASIT AND MARIO TSCHINKE
UNIVERSITY OF PALERMO

COURSE HELD AT THE DEPARTMENT
FOR MECHANICS OF DEFORMABLE BODIES
SEPTEMBER - OCTOBER 1970

UDINE 1970

SPRINGER-VERLAG WIEN GMBH

ISBN 978-3-211-81081-1 ISBN 978-3-7091-2991-3 (eBook)
DOI 10.1007/978-3-7091-2991-3

Originally published by Springer Vienna in 1970

PREFACE

The purpose of lecturing the subject "Photoelasticity in Theory and Practice" at the International Centre for Mechanical Sciences " in the form as presented in this text generated from the idea to give to the participants - who have been in majority well acquainted with the work in a photoelastic laboratory - a view on some basic problems in modern Photoelasticity, without pretending to cover the matter completely.

The physical backgrounds,i.e., the basic equations of electromagnetism, as well as of the optics of anisotropic bodies, and the equations governing the propagation of plane electromagnetic wave in a homogeneous, anisotropic dielectric are given in the first part, as the introduction to the development of photoelastic laws. The general problem of three-dimensional photoelasticity is discussed in more details and a short view on methods to the solution of two-dimensional problem is given.

The second part of the text is dealing with the problem of Photoviscoelasticity giving the attention to the analysis of material properties and to the formulation of photoviscoelastic laws. The problem is discussed considering both the basic theory and the methods of application to engineering problems.

The basic concepts on Photothermoelasticity are the subject of the third part. A special attention is concerned to the problem of photothermo-elastic laws.

The last part deals with some recent problems from the field of Hologram Interferometry applied to Photoelasticity. The basic analysis on Holography, which may be treated as the introduction to this Chapter, has been given in the paper "Application of Holography and Hologram Interferometry to Photoelasticity", published by CISM, Udine, 1969, (see Ref. n. 73). The references 53-71 are related to this paper.

The matter presented in the text was chosen between the numerous existing papers on the matter in question, as shown by reference numbers in the text. Some important fields of modern Photoelasticity e.g., Photoplasticity, Dynamic Photoelasticity, Photoelasticity applied to anisotropic and nonhomogeneous media, Photoelasticity applied to fluids, etc., are not covered by the presented text.

It is a very pleasant obligation to express my gratitude to the Secretary General, Prof. Luigi Sobrero for inviting me to prepare this course for lecturing at the International Centre for Mechanical Sciences in Udine.

September, 1970.

V. BRČIĆ.

Chapter 1.
Basic Theory of Photoelasticity.

1. 1. Introduction.

The term "Photoelasticity" [5]*) describes methods of investigation and numerical evaluation of some mechanical quantities as strain, stress, mode of deformation, velocity distribution, etc..., in solid and fluid bodies by analyzing the change of state and velocity of electromagnetic radiation interacting with these bodies. The major methods of Photoelasticity are based on the analysis of induced and orientational birefringence by measuring the change of polarized radiant power transmitted or reflected from the body under observation.

The detector and carrier of information in photoelastic systems is the photon radiation. The electromagnetic theory of radiation, that can be considered as a particular case of the more general quantum theory, seems to be adequate for formulation of basic and derived theories on which various photoelas-

---

*) [...] References.

ticity techniques are founded. That theory is based on the concept of an electromagnetic field, represented by two vectors, the electric vector and the magnetic vector. The interaction between that field and matter is described by a set of vectors : the electric current density, the electric displacement and the magnetic induction, and by a set of material constants : the specific conductivity, the dielectric constant, and the magnetic permeability.

Since the majority of materials used in photoelasticity is macroscopically homogeneous, practically non-conducting, and isotropic magnetically, in the basic relations of photoelasticity the magnetic permeability is represented by a scalar, and the electric anisotropy is chosen as the most satisfactory optical parameter that alter with strain and stress.

1. 2. Fundamental Equations. [4]

The elastic, isotropic and transparent bodies, which are used in photoelastic investigation and which are submitted to a general three-dimensional state of stresses, behave as birefringent crystals of rhombohedral system, thus, they are submitted to similar mathematical representation. The crystal optics and the theory of electromagnetism are the theoretical background of photoelasticity.

The principles of conservation of charge and magnetic flux lead to the following field equations [8] which have the form:

$$\begin{aligned} \operatorname{div} \underset{\sim}{I} + \frac{\partial Q}{\partial t} &= 0 \\ \operatorname{curl} \underset{\sim}{E} + \frac{\partial \underset{\sim}{B}}{\partial t} &= 0 \\ \operatorname{div} \underset{\sim}{B} &= 0 \end{aligned} \qquad (1.1)$$

where

$\underset{\sim}{B}$ is the density of magnetic flux, or the magnetic induction,

$\underset{\sim}{E}$ is the electric field (or the electric vector),

$\underset{\sim}{I}$ is the current density,

Q is the charge density.

These field equations are related to the existence of electromagnetic potentials such that

$$\begin{aligned} \underset{\sim}{B} &= \operatorname{curl} \underset{\sim}{A} \\ \underset{\sim}{E} &= -\frac{\partial \underset{\sim}{A}}{\partial t} - \operatorname{grad} V \\ Q &= \operatorname{div} \underset{\sim}{D} \\ \underset{\sim}{I} &= \operatorname{curl} \underset{\sim}{H} = \frac{\partial \underset{\sim}{D}}{\partial t} \end{aligned} \qquad (1.2)$$

where

$\underset{\sim}{A}$ is the magnetic potential,

V is the electric potential,

$\underset{\sim}{H}$ is the current potential (or the magnetic vector, or the magnetic field intensity),

$\underset{\sim}{D}$ is the electric displacement (or the charge potential).

The electric vector and the magnetic induction (or the magnetic flux density) are related to the charge potential (or the electric displacement) and the current potential (or the magnetic vector) at all points, inside and outside of the material, by the relations :

$$\begin{aligned} \underset{\sim}{D} &= \varepsilon_0 \underset{\sim}{E} \\ \underset{\sim}{B} &= \mu_0 \underset{\sim}{H} \end{aligned} \tag{1.3}$$

where ε_0 and μ_0 are the dielectric constant (or the specific inductive capacity) and the magnetic permeability for vacuum, respectively. The values ε_0 and μ_0 are fundamental constants depending on the units of time, charge, and magnetic flux.

For vacuum, it holds the well-known relation :

$$c = \frac{1}{\sqrt{\varepsilon_0 \mu_0}}$$

where c is the light velocity in vacuum.

Across a moving surface of discontinuity, such as the boundary between regions of different material properties, which moves with speed u_n and has the unit normal $\mathbf{\nu}$, see Fig. 1, the following discontinuity conditions hold :

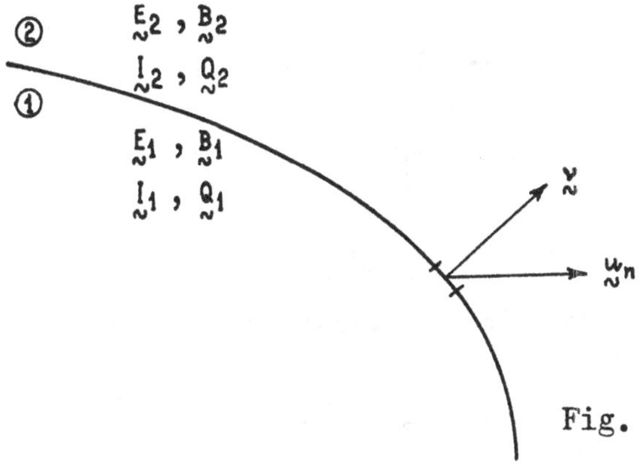

Fig. 1

$$\mathbf{\nu} \times (\mathbf{E}_2 - \mathbf{E}_1) - u_n(\mathbf{B}_2 - \mathbf{B}_1) = 0 \qquad \text{(a)}$$

$$(\mathbf{B}_2 - \mathbf{B}_1)\mathbf{\nu} = 0 \qquad \text{(b)} \qquad (1.4)$$

$$(\mathbf{I}_2 - \mathbf{I}_1)\mathbf{\nu} - u_n(Q_2 - O_1) = 0 \qquad \text{(c)} \; .$$

<u>Constitutive Equations</u>. In addition to the fundamental fields and potentials, constitutive equations defining the <u>electromagnetic nature of the material</u> are required. The formulation of constitutive equations will be given by the introduction of certain

auxiliary fields and potentials :

Let $\underset{\sim}{P}$ be the density of polarization and $\underset{\sim}{M}$ the magnetization field. The polarization current $\underset{\sim}{I}_P$, the bound charge Q_B and the magnetization current $\underset{\sim}{I}_M$ are given by

$$(1.5) \qquad \begin{aligned} \underset{\sim}{I}_P &= \frac{\partial \underset{\sim}{P}}{\partial t} + \mathrm{curl}(\underset{\sim}{P} \times \underset{\sim}{v}) \\ Q_B &= -\mathrm{div}\,\underset{\sim}{P} \\ \underset{\sim}{I}_M &= \mathrm{curl}\,\underset{\sim}{M} \ . \end{aligned}$$

The total charge is considered to be made up of the bound charge plus the free charge :

$$(1.6) \qquad Q = Q_B + Q_F \ .$$

The total current is the sum of the free, the polarization and the magnetization currents :

$$(1.7) \qquad \underset{\sim}{I} = \underset{\sim}{I}_F + \underset{\sim}{I}_P + \underset{\sim}{I}_M \ .$$

We introduce now the partial potentials $\underset{\sim}{\mathcal{G}}$ and $\underset{\sim}{\mathcal{D}}$ defined by the relations

$$(1.8) \qquad \begin{aligned} \underset{\sim}{\mathcal{G}} &= \underset{\sim}{H} - \underset{\sim}{M} - \underset{\sim}{P} \times \underset{\sim}{v} \\ \underset{\sim}{\mathcal{D}} &= \underset{\sim}{D} + \underset{\sim}{P} \ . \end{aligned}$$

The free current and the free charge are then given by

$$\underset{\sim}{I}_F = \operatorname{curl} \underset{\sim}{\mathcal{H}} - \frac{\partial \underset{\sim}{\mathcal{D}}}{\partial t}$$
$$Q_F = \operatorname{div} \underset{\sim}{\mathcal{D}} \quad . \tag{1.9}$$

The discontinuity conditions across a moving surface of discontinuity, with speed u_n and normal $\underset{\sim}{\nu}$, are then :

$$(\underset{\sim}{\mathcal{D}}_2 - \underset{\sim}{\mathcal{D}}_1)\underset{\sim}{\nu} = 0 \tag{1.10}$$

$$\underset{\sim}{\nu} \times (\underset{\sim}{\mathcal{H}}_2 - \underset{\sim}{\mathcal{H}}_1) + u_n(\underset{\sim}{\mathcal{D}}_2 - \underset{\sim}{\mathcal{D}}_1) = 0 \ . \tag{1.11}$$

1. 3. Ideal Dielectrics. [4]

An ideal anisotropic dielectric at rest is defined by the equations :

$$\underset{\sim}{I}_F = 0 \ ; \qquad Q_F = 0$$
$$\underset{\sim}{M} = 0 \ ; \qquad P_i = \varepsilon_0 p_{ij} E_j \ . \tag{1.12}$$

Therefore

$$\mathcal{D}_i = \varepsilon_0 K_{ij} E_j \tag{1.13}$$

where

$$K_{ij} = \delta_{ij} + p_{ij} \tag{1.14}$$

(δ_{ij} is the well-known Kronecker's symbol, and the index notation is used here).

The fundamental equations for dielectrics are :

$$\operatorname{curl} \underset{\sim}{E} + \frac{\partial \underset{\sim}{B}}{\partial t} = 0$$

$$\operatorname{div} \underset{\sim}{B} = 0$$

(1.15)

$$\underset{\sim}{B} = \mu_0 \underset{\sim}{H}$$

$$\mathcal{D}_i = \varepsilon_0 K_{ij} E_j$$

combined with equations

$$\underset{\sim}{I}_F = \operatorname{curl} \mathcal{H} - \frac{\partial \underset{\sim}{\mathcal{D}}}{\partial t}$$

$$Q_F = \operatorname{div} \underset{\sim}{\mathcal{D}}$$

which become

$$\operatorname{curl} \underset{\sim}{\mathcal{H}} - \frac{\partial \underset{\sim}{\mathcal{D}}}{\partial t} = 0$$

(1.16)

$$\operatorname{div} \underset{\sim}{\mathcal{D}} = 0 .$$

The quantity K_{ij} is called <u>the dielectric tensor</u> ; it is the characterization of the optical anisotropy of the considered medium.

1.4. Plane Electromagnetic Wave. [4]

We consider a plane electromagnetic wave propagating through a dielectric at rest. The wave is given by the relations :

$$\underset{\sim}{E} = \underset{\sim}{e}e^{i\varphi} \; ; \quad \underset{\sim}{B} = \underset{\sim}{b}e^{i\varphi} \; ;$$
$$\underset{\sim}{\mathcal{D}} = \underset{\sim}{d}e^{i\varphi} \; ; \quad \underset{\sim}{\mathcal{H}} = \underset{\sim}{h}e^{i\varphi} \; ; \qquad (1.17)$$
$$\varphi = k\underset{\sim}{n}\underset{\sim}{z} - \omega t \; .$$

The unit vector $\underset{\sim}{n}$ gives the direction of the wave ; k is the wave number ; ω is the angular frequency ; $\underset{\sim}{e}$, $\underset{\sim}{b}$, $\underset{\sim}{d}$, and $\underset{\sim}{h}$ are constant, complex vectors.

Substituting Eqs. (1.17) into equations

$$\mathrm{curl}\,\underset{\sim}{E} + \frac{\partial \underset{\sim}{B}}{\partial t} = 0 \; ; \quad \mathrm{curl}\,\underset{\sim}{\mathcal{H}} - \frac{\partial \underset{\sim}{\mathcal{D}}}{\partial t} = 0$$

gives

$$k\underset{\sim}{n} \times \underset{\sim}{e} = \underset{\sim}{b}\omega \qquad \text{(a)}$$
$$k\underset{\sim}{n} \times \underset{\sim}{h} + \omega \underset{\sim}{d} = 0 \qquad \text{(b)} \; . \qquad (1.18)$$

Therefore, $\underset{\sim}{b}$ is perpendicular to $\underset{\sim}{n}$ and $\underset{\sim}{e}$, while $\underset{\sim}{d}$ is perpendicular to $\underset{\sim}{n}$ and $\underset{\sim}{h}$. Equation

$$\underset{\sim}{B} = \mu_0 \underset{\sim}{H}$$

shows that $\underset{\sim}{b}$ is parallel to $\underset{\sim}{h}$, for a dielectric at rest. Thus, $\underset{\sim}{n}$, $\underset{\sim}{d}$, $\underset{\sim}{h}$ are mutually perpendicular, and $\underset{\sim}{d}$, $\underset{\sim}{e}$, $\underset{\sim}{n}$ lie in the same plane (Fig. 2).

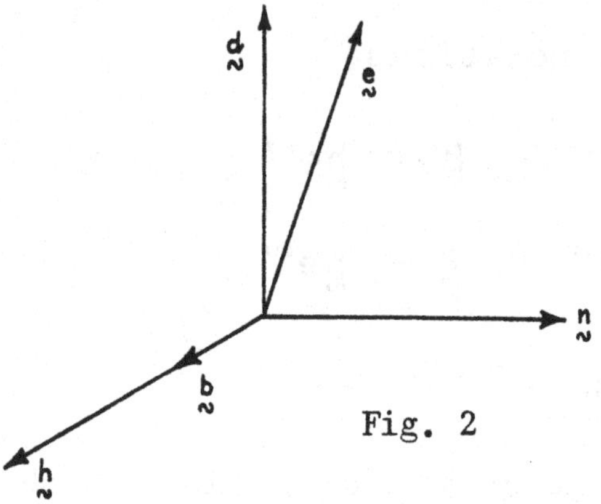

Fig. 2

The plane ($\underset{\sim}{d}$, $\underset{\sim}{h}$) is the wave front.
The direction of $\underset{\sim}{h}$ is the direction of polarization.
The plane ($\underset{\sim}{n}$, $\underset{\sim}{h}$) is the plane of polarization.
The direction of $\underset{\sim}{n}$ is the direction of propagation.
The direction of $\underset{\sim}{e}$ is the direction of vibration.

The constitutive equation for a dielectric,

$$\mathcal{D}_i = \varepsilon_0 K_{ij} E_j$$

gives

$$\varepsilon_0 e_i = K_{ij}^{-1} d_j \ . \tag{1.19}$$

This equation and the second equation in (1.18)

give :

$$e_i = -\frac{k}{\omega\varepsilon_0} K_{ij}^{-1} e_{jrs} n_r h_s \qquad (1.20)$$

where e_{jrs} is the system defined by :

$$e_{jrs} \begin{cases} 0, & \text{when any two of indices are equal,} \\ +1, & \text{when } jrs \text{ is an \underline{even} permutation of } 123, \\ -1, & \text{when } jrs \text{ is an \underline{odd} permutation of } 123, \end{cases}$$

and the index notation is used.

The equation (1.20) and the equations

$$\underset{\sim}{B} = \mu_0 \underset{\sim}{H} \; ; \quad k\underset{\sim}{n} \times \underset{\sim}{e} = \underset{\sim}{b}\omega$$

give

$$(\delta_{ik} + N^2 e_{ipq} e_{jrk} K_{qj}^{-1} n_p n_r) h_k = 0 \qquad (1.21)$$

where

$$N = \frac{kc}{\omega} \qquad (1.22)$$

is the index of refraction, and

$$c = \frac{1}{\sqrt{\varepsilon_0 \mu_0}}$$

is the light velocity.

Let us choose z_i to be the principal axes of K_{ij} and consider the case when $\underset{\sim}{n}$ is directed along z. In this case Eq. (121) becomes :

$$\begin{bmatrix} (1 - \frac{N^2}{K_2}) & 0 & 0 \\ 0 & (1 - \frac{N^2}{K_1}) & 0 \\ 0 & 0 & 1 \end{bmatrix} \begin{Bmatrix} h_1 \\ h_2 \\ h_3 \end{Bmatrix} = \{0\} . \tag{1.23}$$

Eqs. (120) and (123) show that only two plane waves are possible :

Either

$$N = N_1 = \sqrt{K_1} \; ; \; h_1 = h_3 = e_2 = e_3 = 0 \; ; \; h_2 = N_1 c \varepsilon_0 e_1 \tag{1.24}$$

or

$$N = N_2 = \sqrt{K_2} \; ; \; h_2 = h_3 = e_1 = e_3 = 0 \; ; \; h_1 = -N_2 c \varepsilon_0 e_2 \; . \tag{1.25}$$

The direction of polarization coincides with one of the principal axes and the direction of vibration is normal to the wave direction (see Fig 3).

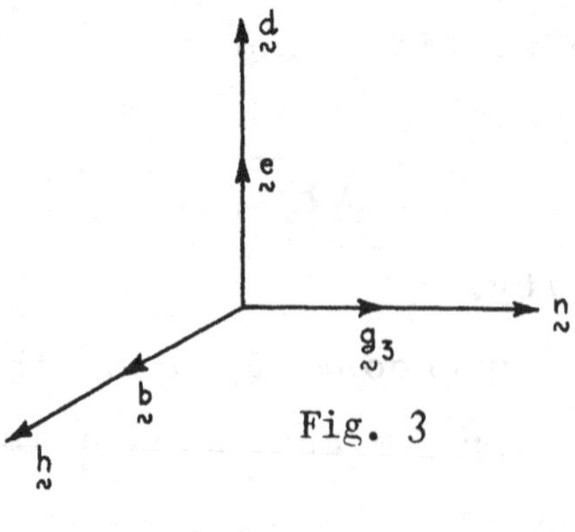

Fig. 3

Every wave must be a linear combination of these two waves.

In free space

$$K_{ij} = \delta_{ij} \text{ , therefore } N = 1.$$

1. 5. Plane Polariscope. [4]

Let consider a thin plate of dielectric bounded by the planes $z = 0$, $z = h$. Consider the case when the coordinate axes are principal axes of the dielectric tensor K_{ij}. By means of a polarizer, a linearly polarized electromagnetic wave propagating along the axis z is obtained. Every such wave can be divided into the sum of monochromatic waves, and every monochromatic wave can be decomposed into components given by Eqs. (1.17), such that the direction of polarization of each component coincides with one of the coordinate axes. Let this plane wave be incident on the surface $z = 0$. In order to satisfy the discontinuity conditions (1.4a) and (1.11), i. e.

$$\underset{\sim}{\nu} \times (\underset{\sim}{E}_{II} - \underset{\sim}{E}_{I}) - u_n(\underset{\sim}{B}_{II} - \underset{\sim}{B}_{I}) = 0$$

$$\underset{\sim}{\nu} \times (\underset{\sim}{\mathcal{H}}_{II} - \underset{\sim}{\mathcal{H}}_{I}) + u_n(\underset{\sim}{\mathcal{D}}_{II} - \underset{\sim}{\mathcal{D}}_{I}) = 0$$

there must also be a wave traveling in the opposite direction [7] (reflected).

The electromametie field preceding the dielectric (see Fig. 4) is therefore the sum of two fields :

(1.26) $$\overset{o}{\underset{\sim}{E}} = \underset{\sim}{E}_{(1)} + \underset{\sim}{E}_{(2)}$$

where

$$\underset{\sim}{E}_{(1)} = \underset{\sim}{e}_{(1)} e^{i(k_0 z - \omega t)} ; \qquad \underset{\sim}{E}_{(2)} = \underset{\sim}{e}_{(2)} e^{i(-k_0 z - \omega t)} .$$

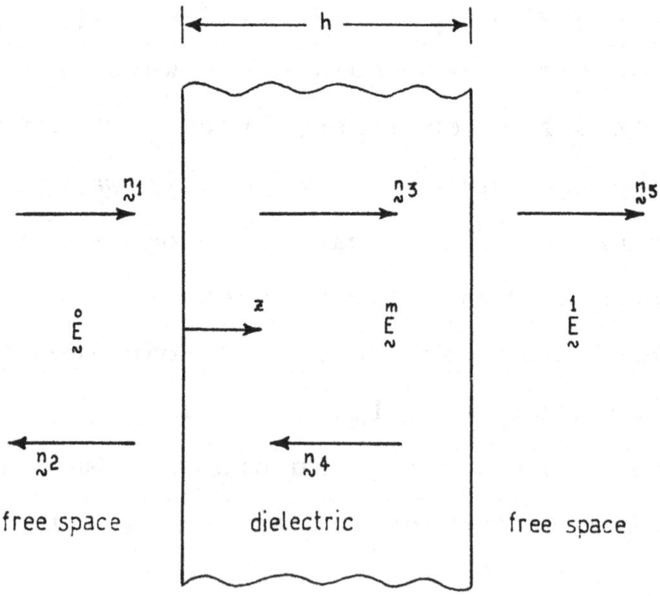

Fig. 4

The wave transmitted into the material is itself reflected at the surface $z = h$. The electromagnetic field in the dielectric material is therefore

$$\overset{m}{\underset{\sim}{E}} = \underset{\sim}{E}_{(3)} + \underset{\sim}{E}_{(4)} \tag{1.27}$$

where

$$\underset{\sim}{E}_{(3)} = \underset{\sim}{e}_{(3)} e^{i(kz - \omega t)} ; \qquad \underset{\sim}{E}_{(4)} = \underset{\sim}{e}_{(4)} e^{i(-kz - \omega t)} .$$

The wave transmitted across the surface $z = h$ gives rise to the electromagnetic field following the dielectric

$$\overset{1}{\underset{\sim}{E}} = \underset{\sim}{e}_{(5)} e^{i(k_0 z - \omega t)} \tag{1.28}$$

where $k_0 = \omega/c$.

The vectors $\underset{\sim}{e}_{(i)}$ are complex constant vector fields which, in the present case, are all directed along the same of the coordinate lines. The value of k used in Eq. (1.27) is determined from the appropriate one of Eqs. (1.24) or (1.25). The corresponding magnetic field is determined from Eq. (1.18a):

$$k \underset{\sim}{n} \times \underset{\sim}{e} = \underset{\sim}{b} \omega .$$

The discontinuity conditions, applied to the plane $z = 0$ and to the plane $z = h$ lead to a set of four vector equations containing the five $\underset{\sim}{e}_{(i)}$

which yield the following result :

$$(1.29) \qquad \underset{\sim}{e}_{(5)} = \frac{4Ne^{-ik_0h}}{(1+N)^2 e^{-ikh} - (1-N)^2 e^{ikh}} \underset{\sim}{e}_{(1)} .$$

If N is of order of magnitude unity, then can be used approximately [1]

$$(1.30) \qquad \underset{\sim}{e}_{(5)} \approx \frac{4N}{(1+N)^2} e^{i(k-k_0)h} \underset{\sim}{e}_{(1)} .$$

Sometimes, the further simplification is used :

$$(1.31) \qquad \underset{\sim}{e}_{(5)} \approx e^{i(k-k_0)h} \underset{\sim}{e}_{(1)} .$$

By assigning the appropriate value to k, this result applies to incident wave polarized along either of the coordinate axes.

Consider an incident monochromatic transverse wave obtained by passing monochromatic light through a polarizer :

$$(1.32) \qquad \underset{\sim}{E}_I = (A \cos\alpha \underset{\sim}{g}_1 + A \sin\alpha \underset{\sim}{g}_2) e^{i(k_0 z - \omega t)}$$

where α is the angle between $\underset{\sim}{E}_I$ and the x-axis.

Applying Eq. (1.29) to each component of the incident wave, the transmitted wave is found to be

$$(1.33) \qquad \underset{\sim}{E}_T = A(c_1 \cos\alpha \underset{\sim}{g}_1 + c_2 \sin\alpha \underset{\sim}{g}_2) e^{i(k_0 z - \omega t)}$$

where

$$c_j = \frac{4N_j e^{-ik_0 h}}{(1 + N_j)^2 e^{-ik_j h} - (1 - N_j)^2 e^{ik_j h}} . \quad (j = 1,2)$$

Passing this wave through a second polarizer, will give a wave of complex amplitude

$$E = A(c_1 \cos\alpha \cos\beta + c_2 \sin\alpha \sin\beta) e^{i(k_0 z - \omega t)} \quad (1.34)$$

where β is the angle between the transmitted wave and the x-axis. This expression can be written in the form :

$$E = 4AK_1 \cos\alpha \cos\beta \, e^{i(k_0 z - \omega t - k_0 h + \theta_1)} + + 4AK_2 \sin\alpha \sin\beta \, e^{i(k_0 z - \omega t - k_0 h + \theta_2)} \quad (1.35)$$

where

$$K_j = \frac{N_j}{2\sqrt{4N_j^2 \cos^2 k_j h + (1 + N_j)^2 \sin^2 k_j h}} \quad (1.36)$$

$$\tan\Theta_j = \frac{1 + N_j^2}{2N_j} \tan k_j h \quad (j = 1, 2) . \quad (1.36a)$$

Either the real or imaginary part of E, Eq. (1.35), may be regarded as the light transmitted.

The imaginary part is

$$I = Q\sin(k_0 z - \omega t - k_0 h - \chi) \tag{1.37}$$

where

$$\begin{aligned} Q\cos\chi &= 4AK_1\cos\alpha\cos\beta\sin\theta_1 + \\ &+ 4AK_2\sin\alpha\sin\beta\sin\theta_2 \ ; \\ Q\sin\chi &= 4AK_1\cos\alpha\cos\beta\cos\theta_1 + \\ &+ 4AK_2\sin\alpha\sin\beta\cos\theta_2 \ . \end{aligned} \tag{1.38}$$

Consider now the case when the first and the second polarizer are crossed, i.e. when $\beta = \alpha + \pi/2$. Then :

$$Q^2 = 4A^2\sin^2 2\alpha(K_1^2 + K_2^2 - 2K_1K_2\cos 2n\pi) \tag{1.39}$$

where the fringe order is defined by

$$n = \frac{\theta_2 - \theta_1}{2\pi} \ . \tag{1.40}$$

Therefore, the light intensity is zero where $\alpha = 0, \pi/2$. The loci of such points are called isoclinic lines. The light intensity will be a local minimum where $N_1/N_2 = c_1, c_2, \ldots$. The loci of such points are called isochromatics.

For the transparent plastics used in photoelasticity N_1 is equal to N_2 in the natural state, and both N_1 and N_2 change very little as a result of the deformation of the material. In such a case, the minimum values of Eq. (1.39) occur at integer values of the fringe order n. If N_1 and N_2 are sufficiently close to unity, Eq. (1.36a) gives:

$$n \approx \frac{h\omega}{2\pi c}(N_2 - N_1) \ . \qquad (1.41)$$

Therefore, for materials with an index of refraction which is near unity and which changes only slightly upon deformation, the isochromatic lines occur at definite values of $N_2 - N_1$ as given by Eq. (1.41).

1. 6. Fresnel's and Index Ellipsoid. [6]

(a) The Fresnel's ellipsoid is defined as the surface of the constant field energy $\underset{\sim}{E}\underset{\sim}{D}$ where the vector $\underset{\sim}{r}(z_1, z_2, z_3)$ of points on the ellipsoid surface is proportional to the electric vector $\underset{\sim}{E}$. In the system of principal axes, it has the equation

$$\frac{z_1^2}{v_1^2} + \frac{z_2^2}{v_2^2} + \frac{z_3^2}{v_3^2} = \frac{1}{c^2} \qquad (1.41a)$$

or

(1.41b) $$\varepsilon_1 z_1^2 + \varepsilon_2 z_2^2 + \varepsilon_3 z_3^2 = \varepsilon_0$$

or (since $n_i^2 = \varepsilon_i \mu$):

(1.41c) $$n_1^2 z_1^2 + n_2^2 z_2^2 + n_3^2 z_3^2 = \mu$$

Therefore, the Fresnel's ellipsoid may be expressed in terms of either velocities , or dielectric constants, or indices of refraction. All these quantities are symmetric tensors of the second order.

The relation between refractive index ($n = c/v$), dielectric constant ε and permeability μ:

(1.42) $$n_i^2 = \varepsilon_i \mu \quad (i = 1,2,3) \; .$$

The relationship between the velocities v_i and the principal dielectric constants ε_i:

(1.43) $$\varepsilon_i = \varepsilon_0 \frac{c^2}{v_i^2} \quad (i = 1,2,3)$$

where ε_0 is the vacuum dielectric constant.

b. <u>The index ellipsoid</u>, or the ellipsoid of the wave normales is defined by the position vector $\underset{\sim}{r}(z_1, z_2, z_3)$ of points on its surface being proportional to the vector $\underset{\sim}{D}$ and this surface is also the surface of the constant field energy $\underset{\sim}{E}\,\underset{\sim}{D}$. In the system of principal axes, this surface is defined by the equation

$$\frac{z_1^2}{\varepsilon_1} + \frac{z_2^2}{\varepsilon_2} + \frac{z_3^2}{\varepsilon_3} = \frac{1}{\varepsilon_0} \qquad (1.44)$$

Since $n_i = c/v_i = \sqrt{\varepsilon_i/\varepsilon_0}$, the principal axes of this ellipsoid are directly proportional to the principal refractive indices.

Note : The electric energy per unit volume is equal to

$$\frac{1}{8\pi}(\varepsilon_1 E_1^2 + \varepsilon_2 E_2^2 + \varepsilon_3 E_3^2) \qquad \text{(in principal coordinates)}$$

or

$$\frac{1}{8\pi}(\varepsilon_{11} E_1^2 + \varepsilon_{22} E_2^2 + \varepsilon_{33} E_3^2 + 2\varepsilon_{12} E_1 E_2 + 2\varepsilon_{23} E_2 E_3 + 2\varepsilon_{31} E_3 E_1) \; . \qquad (1.45)$$

It may be assumed that in the first approximation the changes in the components of the dielectric tensor or of the index tensor are homogeneous linear functions of the six stress or strain components. Consequently, the electric vector of the electromagnetic radiation may be considered as the detector and carrier of information on stress or strain components.

1. 7. The Photoelastic Laws. [2]

Under the action of external load the homogeneous, isotropic transparent plastics, used in photoelasticity, become birefringent and behave as an anisotropic homogeneous body, f.i. a crystal. This effect, called the stress-optical effect, was studied by F.E. Neumann and J.C. Maxwell. They assumed that the effect was due to changes in the molecular arrangement of the molecules produced by load.

Neumann (1841) accepted the strain-optical description : the difference of velocities of two oppositely polarized waves in light passing through the tested beam (in flexure) in a direction normal to the plane of bending was directly proportional to the difference of the two principal strains in the plane of the wavefront.

Therefore, he assumed that the three principal velocities v_1, v_2, v_3, were expressed by the principal strains, i.e.(the alignment of the principal axes has been assumed) :

$$\begin{aligned} v_1 &= v_0 + \alpha\varepsilon_1 + \beta\varepsilon_2 + \gamma\varepsilon_3 \\ v_2 &= v_0 + \alpha\varepsilon_2 + \beta\varepsilon_3 + \gamma\varepsilon_1 \end{aligned} \quad (1.46a)$$

$$v_3 = v_0 + \alpha\varepsilon_3 + \beta\varepsilon_1 + \gamma\varepsilon_2 \tag{1.46b}$$

where v_0 is the velocity in unstrained medium. From (1.46) :

$$v_1 - v_2 = (\alpha - \gamma)\varepsilon_1 + (\beta - \alpha)\varepsilon_2 + (\gamma - \beta)\varepsilon_3 \ .$$

It was experimentally shown that the difference $v_1 - v_2$ may be considered as independent on ε_3, therefore $\gamma = \beta$ and

$$v_1 - v_2 = (\alpha - \beta)(\varepsilon_1 - \varepsilon_2) \ . \tag{1.47}$$

Thus,

$$\begin{aligned} v_1 &= v_0 + \alpha\varepsilon_1 + \beta(\varepsilon_2 + \varepsilon_3) \\ v_2 &= v_0 + \alpha\varepsilon_2 + \beta(\varepsilon_3 + \varepsilon_1) \\ v_3 &= v_0 + \alpha\varepsilon_3 + \beta(\varepsilon_1 + \varepsilon_2) \ . \end{aligned} \tag{1.48}$$

Maxwell (1853) expressed the principal wave velocities in terms of stresses :

$$\begin{aligned} v_1 &= v_0 + C_1\sigma_1 + C_2(\sigma_2 + \sigma_3) \\ v_2 &= v_0 + C_1\sigma_2 + C_2(\sigma_3 + \sigma_1) \\ v_3 &= v_0 + C_1\sigma_3 + C_2(\sigma_1 + \sigma_2) \ . \end{aligned} \tag{1.49}$$

α and β are called strain optical constants (or elasto-optical constants).

C_1 and C_2 are called stress-optical constants (or piezo-optical constants).

From Eqs. (1.49) it follows that the difference of principal velocities is

$$v_1 - v_2 = (C_1 - C_2)(\sigma_1 - \sigma_2) . \tag{1.50}$$

For the application to perfectly elastic materials both formulations, strain- and stress-optical, are equivalent. In other cases, f.i. at visco elastic materials, it is necessary to introduce the corresponding constants according to either strain or stress formulation.

Usually, it is more convenient to work in terms of stress-optical coefficients.

By comparing the properties of index ellipsoid and the stress quadric, we can deduce from the consideration of symmetry that the two surfaces must have the same principal axes. These equations have the form :

$$\begin{aligned} v_1^2 z_1^2 + v_2^2 z_2^2 + v_3^2 z_3^2 &= 1 \\ \sigma_1 z_1^2 + \sigma_2 z_2^2 + \sigma_3 z_3^2 &= 1 . \end{aligned} \tag{1.51}$$

The squares of the principal wave velocities are related here to the principal stresses and it may be taken that it may be the squares of the velocities rather then the velocities themselves which should be the linear functions of the stresses. Thus :

$$\begin{aligned} v_1^2 &= v_0^2 + C_1'\sigma_1 + C_2'(\sigma_2 + \sigma_3) \\ v_2^2 &= v_0^2 + C_1'\sigma_2 + C_2'(\sigma_3 + \sigma_1) \qquad (1.52) \\ v_3^2 &= v_0^2 + C_1'\sigma_3 + C_2'(\sigma_1 + \sigma_2) \, . \end{aligned}$$

If Maxwell's equations (1.49) are squared they give :

$$\begin{aligned} v_1^2 &= v_0^2 + 2v_0C_1\sigma_1 + 2v_0C_2(\sigma_2 + \sigma_3) + \{C_1\sigma_1 + C_2(\sigma_2 + \sigma_3)\}^2 \\ v_2^2 &= v_0^2 + 2v_0C_1\sigma_2 + 2v_0C_2(\sigma_3 + \sigma_1) + \{C_1\sigma_2 + C_2(\sigma_3 + \sigma_1)\}^2 \\ v_3^2 &= v_0^2 + 2v_0C_1\sigma_3 + 2v_0C_2(\sigma_1 + \sigma_2) + \{C_1\sigma_3 + C_2(\sigma_1 + \sigma_2)\}^2 . \end{aligned}$$

After neglecting terms of the second order in the small coefficients C_1 and C_2 the last equation reduces to the form (1.52), where

$$C_1' = 2v_0C_1 \, ; \qquad C_2' = 2v_0C_2 \, . \qquad (1.53)$$

Eqs. (1.52) can be written more conveniently :

$$v_1^2 = v_0^2 + C_2'(\sigma_1 + \sigma_2 + \sigma_3) + (C_1' - C_2')\sigma_1 \qquad (1.54a)$$

$$(1.54b)\quad \begin{aligned} v_2^2 &= v_0^2 + C_2'(\sigma_1 + \sigma_2 + \sigma_3) + (C_1' - C_2')\sigma_2 \\ v_3^2 &= v_0^2 + C_2'(\sigma_1 + \sigma_2 + \sigma_3) + (C_1' - C_2')\sigma_3 \,. \end{aligned}$$

After substituting the values for v_1^2, v_2^2, v_3^2 into the equation of index ellipsoid we get :

$$\{v_0^2 + C_2'(\sigma_1 + \sigma_2 + \sigma_3)\}(z_1^2 + z_2^2 + z_3^2) + \\ + (C_1' - C_2')(z_1^2\sigma_1 + z_2^2\sigma_2 + z_3^2\sigma_3) = 1$$

or

$$(1.55)\quad \frac{1}{R^2} = v_0^2 + C_2'(\sigma_1 + \sigma_2 + \sigma_3) + \\ + (C_1' - C_2')(\ell^2\sigma_1 + m^2\sigma_2 + n^2\sigma_3)$$

where R is the radius vector from the centre to the ellipsoid surface in the direction (ℓ, m, n).

The maximum and minimum values of R in (1.55) correspond to maximum and minimum values of the expression $(\ell^2\sigma_1 + m^2\sigma_2 + n^2\sigma_3)$ (because $\sigma_1 + \sigma_2 + \sigma_3$ is the first stress invariant) and this is the expression for the normal stress in direction (ℓ, m, n). We can deduce that the directions of polarization in any wave front are the directions of the secondary principal stresses in the plane of the wave-front (since

R_{max} and R_{min} correspond to minimum and maximum values of the normal stress in the considered central section of the index ellipsoid).

This is the first Maxwell's photoelastic law.

If the secondary principal stresses in a given wavefront are σ_p and σ_q and R_p and R_q the lengths of the radii-vectors in their directions, then

$$\left.\begin{aligned} \frac{1}{R_p^2} &= v_0^2 + C_2'(\sigma_1 + \sigma_2 + \sigma_3) + (C_1' + C_2')\sigma_p \\ \frac{1}{R_q^2} &= v_0^2 + C_2'(\sigma_1 + \sigma_2 + \sigma_3) + (C_1' + C_2')\sigma_q \end{aligned}\right\} \qquad (1.56)$$

and we have :

$$\frac{1}{R_p^2} - \frac{1}{R_q^2} = (C_1' - C_2')(\sigma_p - \sigma_q) \, . \qquad (1.57)$$

But in index ellipsoid

$$\frac{1}{R_p} = v_p \, ; \quad \frac{1}{R_q} = v_q \, .$$

Thus,

$$\frac{1}{R_p^2} - \frac{1}{R_q^2} = v_p^2 - v_q^2 \approx 2v_0(v_p - v_q)$$

or

$$v_p - v_q = C(\sigma_p - \sigma_q) \, . \qquad (1.58)$$

This is the second Maxwell's photoelastic law : The difference of the velocities of the two oppositely polarized waves in any wave-front is proportional to the difference of the (secondary) principal stresses in the plane of the wave-front.

However, if the first law is exact, the exact form of the second law should be :

The difference of the squares of the velocities of the two oppositely polarized waves is proportional to the difference of the (secondary) principal stresses in the plane of the wave-front.

Thus, when the model is stressed, it divides each entering ray into two plane-polarized components, polarized in the directions of principal stresses, which propagate through the model with different velocities.

In the case of two-dimensional state of stresses we have the principal (no secondary principal) stresses, and Eq. (1.58) has the form :

$$v_1 - v_2 = C(\sigma_1 - \sigma_2) \quad (1.59)$$

where σ_1 and σ_2 are the principal stresses.

1. 8. Photoelastic System.

The purpose of the arrangement called a photoelastic system or a photoelastic polariscope is to detect, to measure, and to record the modulation of the flow of the radiant energy by the stress and strain states in birefringent bodies [5]. That modulation results in the well-known change of the velocity of radiation what results in the phase retardation and in the change of the mode of propagation.

The flow pattern of the radiant energy modulated by the strain and stress components in a photoelastic body are transformed by the polariscope elements into changes of the radiation power which can be easier detected, measured, and recorded.

The majority of polariscopes are designed for detecting and measuring relative birefringence. However, recently developed interferometer methods offer the possibility of measuring absolute retardation.

The typical photoelastic system for measuring relative retardation consists of the following elements : radiation source, optical system, polarizing units (polarizer and retarders), detector and/or recorder, photoelastic object, and image forming system.

In the following, the considerations will be restricted to the analysis of a parallel radiation beam through a photoelastic system consisting of plane elements which are normal to the direction of energy flow. The considerations relate to steady-state problems, without changing the frequency of radiation and temperature.

1.8.1. Passage of the Plane Polarized Light Through a Stressed Plate [2].

Consider a beam of plane polarized light with vibrations in direction $0x$, incident upon a flat parallel plate to the edges of which are applied forces acting parallel to the faces of the plate. Let $0\sigma_1$ and $0\sigma_2$ be the directions of principal stresses at a point 0 of the plate (Fig 5).

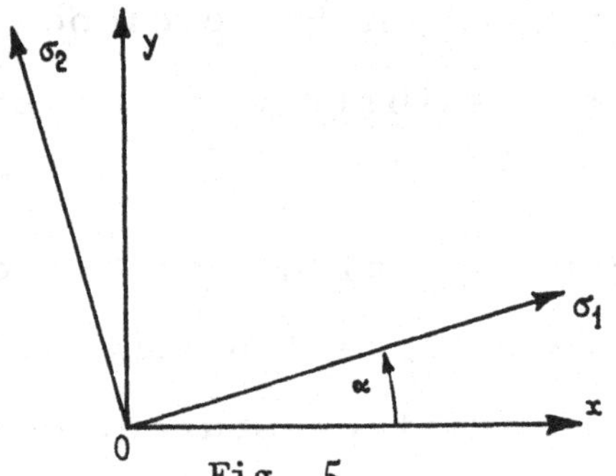

Fig. 5

If the original plane-polarized wave is

$$u_x = a\cos\omega t \tag{1.60}$$

it will be polarized on entering the plate at 0 into

$$u_{\sigma_1} = a\cos\alpha\cos\omega t\;; \qquad u_{\sigma_2} = -a\sin\alpha\cos\omega t \;.$$

These waves travel with velocities v_1 and v_2 , and will emerge after traveling through the thickness d of the plate as

$$u'_{\sigma_1} = a\cos\alpha\cos\omega\left(t - \frac{d}{v_1}\right)$$

$$u'_{\sigma_2} = -a\sin\alpha\cos\omega\left(t - \frac{d}{v_2}\right) .$$

If the light now passes through the analyzer "crossed" with the polarizer, only the components of these vibrations parallel to Oy will emerge giving the vibration

$$u_y = a\sin2\alpha\sin\omega\left(\frac{d}{2v_1} - \frac{d}{2v_2}\right)\sin\omega\left(t - \frac{d}{2v_1} - \frac{d}{2v_2}\right) . \quad (1.61)$$

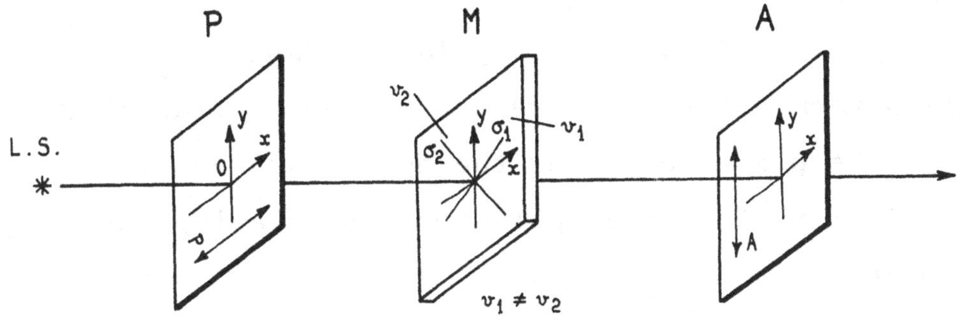

Fig. 6

In the case, the light is monochromatic, the intensity of the transmitted light will be proportional to the square of the amplitude, i.e. to

$$\sin^2 2\alpha\,\sin^2\omega\left(\frac{d}{2v_1} - \frac{d}{2v_2}\right) . \quad (1.62)$$

Considering the first factor of this

expression, we see that the intensity will vary from point to point of the plate with the variation in the directions of the principal stresses. It will be equal to zero for $\alpha = 0$, $\alpha = 90°$, maximum at $\alpha = 45°$. The dark areas have the appearence of fringes which are called isoclinic fringes (loci of points in the plate at which the principal stresses are parallel to the vibration directions of polarizer and analyzer). If the inclination of polarizer and analyzer axes to the arbitrarily chosen reference x-direction be φ, the lines observed are referred to as isoclinic lines of parameter φ.

If we rotate polarizer and analyzer and analyzer keeping them crossed, these lines move continuously across the plate, enabling the determination of the directions of the principal stresses at all points of the plate.

The intensity of the light will be equal to zero when

$$(1.63) \qquad \omega\left(\frac{d}{2v_1} - \frac{d}{2v_2}\right) = 2\pi\frac{c}{\lambda}\left(\frac{d}{2v_1} - \frac{d}{2v_2}\right) = \begin{cases} 0 \\ i\pi \end{cases}$$

or $\quad d(n_1 - n_2) = i\lambda \qquad (i = 0, 1, 2, \ldots)$

where n_1 and n_2 are the indices of refraction for the two oppositely polarized waves.

The expression

$$d(n_1 - n_2) \qquad \text{or} \qquad d\left(\frac{c}{v_1} - \frac{c}{v_2}\right)$$

gives the difference in "optical paths" of the two waves ("relative retardation"). But

$$d\left(\frac{c}{v_1} - \frac{c}{v_2}\right) = \frac{cd}{v_1 v_2}(v_2 - v_1) \approx \frac{cd}{v_0^2}(v_2 - v_1) \cong (\sigma_1 - \sigma_2)$$

where v_0 is the light velocity in the unstressed medium. Thus, we have finally :

$$\text{Relative retardation} = i\lambda = Cd(\sigma_1 - \sigma_2) \qquad (1.64)$$

where $i = 0, 1, 2, \ldots,$ and C is a stress-optical coefficient.

Eq. (1.64) gives the isochromatics (the loci of points of the constant principal stress differences) of the order i, and the set of the isochromatics at a specific load represents the stress fringe pattern.

The fringe value of the material f represents the stress-difference in kg/cm^2 which produces a relative retardation of one wave length in light passing through the one centimeter thickness of the model.

$(\sigma_1 - \sigma_2)$ in kg/cm^2 = f x relative retardation in wave length/thickness in cm.

Finally :

$$\sigma_1 - \sigma_2 = i\frac{f_\sigma}{d} \qquad (1.64')$$

where i is isochromatic order, f_σ model constant, and d model thickness.

1.8.2. Passage of Circularly Polarized Light Through a Stressed Model.

The equipment arrangement is represented schematically in Fig. 7.

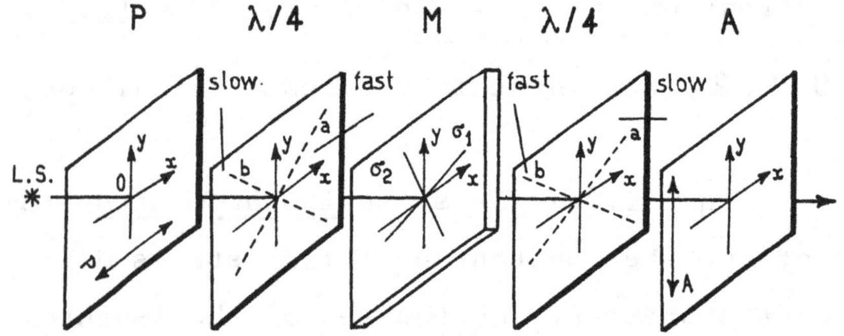

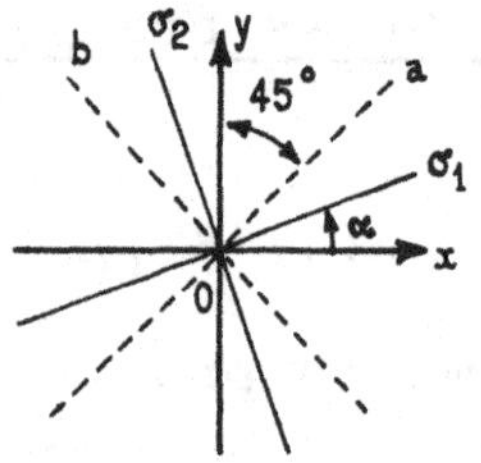

Fig. 7

In traveling through the quarter wave plate, one wave will be retarded on the other by $\lambda/4$ or will suffer a phase retardation of $\pi/4$. At the first $\lambda/4$ plate let Oa be the fast, and at the second $\lambda/4$ plate, Ob the fast axis. The quarter waves plates are "crossed", polarizer and analyzer also crossed.

The final amplitude of the light (after passing analyzer) :

$$u_y = \frac{\sqrt{2}}{2} \sin\omega\left(\frac{d}{2v_2} - \frac{d}{2v_1}\right)\cos(\omega t - \beta) \, . \qquad (1.65)$$

This expression will be equal to zero when

$$\omega\left(\frac{d}{2v_2} - \frac{d}{2v_1}\right) = i\pi \qquad (i = 0, 1, 2, \ldots,) \, . \, (1.66)$$

In monochromatic light, the field of the polariscope will be dark, with dark fringes, as in the case of the plane-polariscope, but now, there are no isoclinic fringes.

Two basic possibilities of the polariscope arrangement :

(1) Dark field and whole wave-length fringes given by::

  (a) P and A crossed, and $\lambda/4$ plates crossed,

  (b) P and A parallel, and $\lambda/4$ plates parallel.

(2) Light field and odd half wave-length fringes given by :

  (a) P and A crossed, and $\lambda/4$ plates parallel,

  (b) P and A parallel, and $\lambda/4$ plates crossed.

1. 9. Propagation of Electromagnetic Waves in a Nonhomogeneous, Anisotropic Medium. General Equations of Three-Dimensional Photoelasticity.

The basic problem in three-dimensional photoelasticity is to find suitable solution of Maxwell's equations which govern to sufficient approximation the propagation of electromagnetic waves in weakly anisotropic and nonhomogeneous media. There is an extensive literature on the homogeneous, anisotropic media, and the nonhomogeneous, isotropic media, but the combined problem was attacked in its most general form by only few authors.

The first trial, based on purely kinematical considerations, is originated by F.E. Neumann (1841). Later, a series of papers (R.D. Mindlin and L.E. Goodman [9], R.C. O'Rourke [10], D.C. Drucker and W.B. Woodward [11], H. Aben [12], [13], [14]) appeared trying to give the suitable form of corresponding differential equations with the purpose of possible practical applications.

Neumann [15] considered the effect of variation of stress through the thickness and he gave the approximative solution for the case when the variation of the directions of principal stresses through the thickness is small.

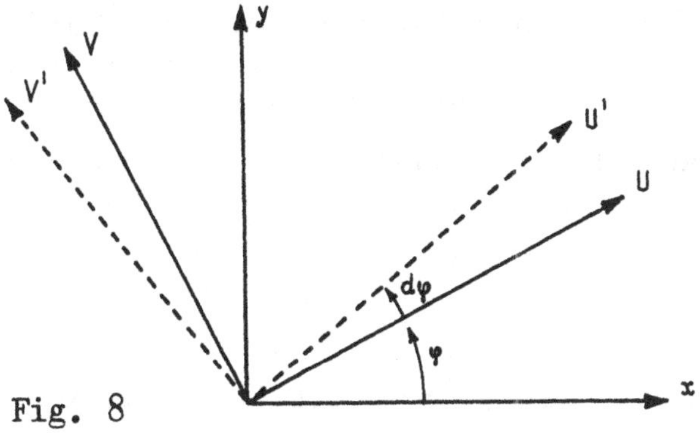

Fig. 8

Let be (see Fig. 8) OU and OV the directions of principal stresses in a layer of the considered body, and the components of the light vector along these directions :

$$u = a\sin(\omega t - \varepsilon_1)\ ; \qquad v = b\sin(\omega t - \varepsilon_2)\ .$$

Here, ε_1 and ε_2 are phase retardations in radians. At the distance dz the components of the light vector are :

$$u = a\sin(\omega t - \varepsilon_1 - \frac{2\pi}{\lambda} n_1 dz)$$

$$v = b\sin(\omega t - \varepsilon_2 - \frac{2\pi}{\lambda} n_2 dz)$$

where n_1 and n_2 are the refractive indices of the two waves at the considered location. The variation of the angle of the principal axes through the thickness dz let be equal to $d\varphi$.

Neglecting the changes of light intensity in passing through the layer and neglecting small quantities of higher orders, Neumann obtained, by using purely kinematical considerations, two governing equations :

$$
\begin{aligned}
d\gamma &= -\cos(\varepsilon_1 - \varepsilon_2)d\varphi \\
d\varepsilon_1 - d\varepsilon_2 &= 2\,\mathrm{ctan}\,2\gamma \sin(\varepsilon_1 - \varepsilon_2)d\varphi + \frac{2\pi}{\lambda}(n_1 - n_2)dz
\end{aligned}
\tag{1.67}
$$

where $\gamma = \arctan(b/a)$.

Neumann discussed the possibility of the application to practical problems using the solution of those equations.

R.D. Mindlin and L.E. Goodman [9] gave the solution for the case where the principal axes of the section of the index ellipsoid vary linearly along the wave normal. Later, Mindlin, and Drucker generalized the problem considering the case when the rotation of principal axes is a linear function of the distance along the wave normal.

Mindlin showed that the passage from the dynamical Maxwell's equations to the kinematical Neumann's equations is possible. He started with

Maxwell's equations

$$\dot{\underset{\sim}{D}} = \mathrm{curl}\,\underset{\sim}{H} \quad (a) \qquad \underset{\sim}{B} = \mu \underset{\sim}{H} \quad (c)$$
$$\dot{\underset{\sim}{B}} = -\,\mathrm{curl}\,\underset{\sim}{E} \quad (b) \qquad D_i = \varepsilon_{ij} E_j \quad (d) \qquad (1.68)$$

where μ is constant magnetic permeability, and the heterogeneity and anisotropy of the medium are characterized by the tensor of dielectric constants ε_{ij} whose elements are the reciprocals of the squares of the indices of refraction n_{ij}. If $\underset{\sim}{r}$ is the position vector, the index ellipsoid is given by the matrix equation

$$\{r\}'[\varepsilon]\,\{r\} = 1 \, . \qquad (1.69)$$

In an elastic isotropic three-dimensional medium the general form of the stress-optical relation may be written as

$$\varepsilon_{ij} = N_0^2 \delta_{ij} + C_0 \sigma_{ij} + C_1 \sigma_{kk} \delta_{ij} \qquad (1.70)$$

where C_0 and C_1 are stress-optical constants, and δ_{ij} is the Kronecker's symbol.

By expressing Eqs. (1.68) by the magnetic vector (i.e. by combining Eqs. 1.68a, 1.68b, 1.68c) one obtains :

$$\mu \ddot{\underset{\sim}{H}} = -\,\mathrm{curl}\,\{[\varepsilon]\,\mathrm{curl}\,\underset{\sim}{H}\} \, ; \; \mathrm{div}\,\underset{\sim}{H} = 0 \, . \qquad (1.71)$$

Eqs. (1.71), with $[\varepsilon]$ given by (1.70), are the dynamical equations governing the propagation of light in the considered medium.

In the case of three-dimensional state of stresses the analysis of the general solution of Eq. (1.71) leads, after some simplifications, to the conclusion that the propagation of the electromagnetic wave in z-direction in a photoelastic body is governed by the differential equations,

$$
(1.72) \qquad \begin{aligned}
\frac{\partial u'}{\partial z} - v'\frac{d\varphi}{dz} &= -\frac{1}{k_2}\frac{\partial u'}{\partial t} \\
\frac{\partial v'}{\partial z} + u'\frac{d\varphi}{dz} &= -\frac{1}{k_1}\frac{\partial v'}{\partial t}
\end{aligned}
$$

where

$$
u = H_x \;; \qquad v = H_y \;;
$$

$$
u' = H_{s_1} \;; \qquad v' = H_{s_2}
$$

(the components of $\underset{\sim}{H}$ in the directions of the secondary principal stresses), and φ is the angle between x and s_1. k_1 and k_2 are the velocities corresponding to n_1 and n_2 and

$$
k_1^2 = \frac{1}{\mu_1 n_1^2} \;; \qquad k_2^2 = \frac{1}{\mu_2 n_2^2} \,.
$$

Equations (1.72) have in two-dimensional case the form :

$$\frac{\partial u'}{\partial z} = \mp \frac{1}{b} \frac{\partial u'}{\partial t} \;; \qquad \frac{\partial v'}{\partial z} = \mp \frac{1}{a} \frac{\partial v'}{\partial t} \qquad (1.73)$$

and they lead to the usual photoelastic laws, namely, the directions of polarization coincide with the directions of principal stresses in the plane perpendicular to the wave normal. The relative retardation is

$$\delta = \omega z \left(\frac{1}{b} - \frac{1}{a}\right) = \frac{2\pi z}{\lambda}(n_b - n_a) \qquad (1.74)$$

where a and b are the principal velocities given by the expressions :

$$a^2 = \frac{1}{\mu n_a^2} \;; \qquad b^2 = \frac{1}{\mu n_b^2} \qquad (1.75)$$

n_a and n_b are the principal refractive indices.

Lately Hillar K. Aben [12], [13], [14] published a series of papers treating the general problem of three-dimensional photoelasticity analyzing the propagation of electromagnetic waves in nonhomogeneous, anisotropic medium and the corresponding solution of Maxwell's equations. As he mentioned, the theoretical aspect of this problem is connected more closely with electrodynamics of plasmas than with crystal optics on which the majority of photoelastic methods are based. The Aben's idea is to investigate the possibilities of working out a nondestructive method of three-dimensional photoelasticity which would be based on integral optical phenomena and which would take into account the rotation of principal axes.

Aben started with a system of differential equations proposed by V.L. Ginsburg [16] (1941) governing the propagation of electromagnetic waves in nonhomogeneous, anisotropic medium :

$$(1.76)\qquad \begin{aligned} \frac{d^2E_1}{dz^2} + \frac{\omega^2}{c^2}D_1 &= 0 \\ \frac{d^2E_1}{dz^2} + \frac{\omega^2}{c^2}D_2 &= 0 \end{aligned}$$

where E_i and D_i ($i = 1, 2$) are the components of the electric vector $\underset{\sim}{E}$ and of the electric induction (displacement) $\underset{\sim}{D}$ in an arbitrary coordinate system. The direction of the wave normal coincides with the z-axis.

By using the expression

$$(1.77)\qquad D_i = \varepsilon_{ij}E_j \qquad\qquad (i = 1, 2)$$

and the solution of Eqs. (1.76) in the form

$$E_j = A_j e^{-ikz} \qquad \left(k = \frac{\omega N}{c}\,;\; j = 1,2\right) \qquad (1.78)$$

where N is the index of refraction, we obtain after neglecting small order terms :

$$\begin{aligned} \frac{dB_1}{dz} &= -\frac{1}{2} i C(\varepsilon_{11} - \varepsilon_{22})B_1 - iC\varepsilon_{12}B_2 \\ \frac{dB_2}{dz} &= -iC\varepsilon_{21}B_1 + \frac{1}{2} iC(\varepsilon_{11} + \varepsilon_{22})B_2 \end{aligned} \qquad (1.79)$$

where

$$\begin{aligned} A_j &= B_j e^{if(z)} \\ f(z) &= \frac{1}{2}kz - \frac{1}{2}C\int(\varepsilon_{11} + \varepsilon_{22})dz \\ C &= \frac{1}{2k}\frac{\omega^2}{c^2} \;. \end{aligned} \qquad (1.80)$$

Eqs. (1.79) can be considered as general equations which govern the propagation of electromagnetic waves in a weakly anisotropic, nonhomogeneous medium. Using the transformation to the rotated system of principal axes

$$B_1' = B_1\cos\varphi + B_2\sin\varphi \qquad (1.81a)$$

(1.81b) $$B_2' = -B_1 \sin\varphi + B_2 \cos\varphi$$

and the well-known equation (being valid for the case of elastic deformations)

(1.82) $$\varepsilon_{ij} = N^2\delta_{ij} + C_0\sigma_{ij} + C_1\sigma_{kk}\delta_{ij}$$

we get finally :

(1.83) $$\begin{aligned} \frac{dB_1}{dz} &= -\frac{1}{2} i C'(\sigma_{11} - \sigma_{22})B_1 - iC'\sigma_{12}B_2 \\ \frac{dB_2}{dz} &= -iC'\sigma_{21}B_1 + \frac{1}{2} iC'(\sigma_{11} - \sigma_{22})B_2 \end{aligned}$$

and

(1.84) $$\begin{aligned} \frac{dB_1'}{dz} &= -\frac{1}{2} C'(\sigma_1 - \sigma_2)B_1' + \frac{d\varphi}{dz} B_2' \\ \frac{dB_2'}{dz} &= -\frac{d\varphi}{dz} B_1' + \frac{1}{2} iC'(\sigma_1 - \sigma_2)B_2' \end{aligned}$$

where $C' = CC_0$.

Equations (1.83) and (1.84) may be considered as general equations of three-dimensional photoelasticity. These equations contain as special cases the equations proposed by Ginsburg, Drucker, Mindlin and Goodman, and they are equivalent to the nonlinear Neumann's equations.

The question which arises now is to

determine which experimental data can be obtained if the rotation of principal axes is present and how to use these data in order to determine the state of stresses.

If the ratio of rotation of principal axes to the phase retardation is small, the incident light which is linearly polarized along a principal axis at the point of entrance will follow approximately the principal axes and emerge practically linearly polarized. However, in many cases this approximation can lead to significant errors.

In the case of arbitrary rotation of principal axes, as it was shown by H. Aben, there always exist two perpendicular directions of the polarizer by which the light emerging from the model is linearly polarized. Although these polarization directions, in general do not coincide with the axes of principal stresses, their experimental determination gives additional informations about the stresses in the model.

Aben proposed the solution of the system (1.83) or (1.84) in the form

$$\begin{Bmatrix} B_1 \\ B_2 \end{Bmatrix} = [u] \begin{Bmatrix} B_{10} \\ B_{20} \end{Bmatrix} \qquad (1.85)$$

where B_{10} and B_{20} are the components of the incident light vector, B_1 and B_2 the components of the emergent light vector, and the matrix $[U]$, denoting a linear transformation, is a unitary matrix having the form

$$[U] = \begin{bmatrix} e^{i\xi}\cos\vartheta & e^{i\zeta}\sin\vartheta \\ -e^{-i\zeta}\sin\vartheta & e^{-i\xi}\cos\vartheta \end{bmatrix} \tag{1.86}$$

where parameters ξ , ζ and ϑ are determinable by the state of stresses on the wave normal between the points of entrance and emergence of light.

Aben introduces the angles α_1 and α_2 showing the rotation of the coordinate axes at the point of entrance and emergence, respectively, relating them to the matrix $[U]$, i.e. to the parameters ξ , ζ, ϑ . Directions determined by α_1 and α_2 are called <u>primary and secondary characteristic directions</u> of the photoelastic medium. It can be shown, if the incident light is linearly polarized in one of the two perpendicular primary characteristic directions of the considered medium, then the emergent light is also linearly polarized in the corresponding secondary characteristic directions.

The characteristic directions can be determined experimentally by transmission polariscope if the polarizer and analyzer can be rotated separately (looking for the complete extinction of the light). The characteristic phase retardation can be obtained by ordinary methods treating characteristic directions as principal directions. However, generally, the characteristic directions do not coincide with principal directions at the point of entrance and emergence of the light (see Reference 13).

Characteristic quantities α_1 , α_2 , adding one more quantity, γ , the characteristic phase retardation, allow the determination of the parameters ξ , ζ , ϑ , which are functions of the state of stresses on the light path. Thus, experimental determination of characteristic quantities gives, in general, on every light path the relations determining the stress components.

The described Aben's method gives very good results in some particular cases (thin shells, axially symmetrical problems, multilayer media), and it can be combined with some other method too.

1. 10. The Methods to the Solution of Plane Problem.

Three data are necessary to the solution of the plane problem of Elasticity. By using two-dimensional photoelasticity and normal incidence method, we are able to get two data :

(1) the angle α of the principal stress σ_1.

(2) the difference of the principal stresses $\sigma_1 - \sigma_2$.

One more information is still necessary. The problem consists in obtaining either separate values of principal stresses, or the values of the stress components σ_x, σ_y, and τ_{xy}. In the following text, a short view on mostly used methods relating the problem of separation of principal stresses will be given.

1.10.1. Integration of the Lamé-Maxwell's Differential Equations.

The Lamé-Maxwell's differential equations

$$\frac{\partial \sigma_1}{\partial s_1} + \frac{\sigma_1 - \sigma_2}{\varrho_2} = 0 ; \quad \frac{\partial \sigma_2}{\partial s_2} + \frac{\sigma_1 - \sigma_2}{\varrho_1} = 0$$

become after integrating along stress trajectories

$$(1.87) \quad \begin{aligned} \sigma_1 &= (\sigma_1)_0 - \int_0^s (\sigma_1 - \sigma_2)\frac{ds_1}{\varrho_2} \\ \sigma_2 &= (\sigma_2)_0 - \int_0^s (\sigma_1 - \sigma_2)\frac{ds_2}{\varrho_1} \end{aligned}$$

At a free boundary the direct solution is obtainable because one of the two principal stresses is equal to zero. Therefore, using the isochromatic pattern the constants $(\sigma_1)_0$ and $(\sigma_2)_0$ can be obtained directly and the problem is deduced to the numerical integration along trajectories.

1.10.2. Shear Difference Method.

The equilibrium equations are used now (usually neglecting body forces):

$$\frac{\partial \sigma_x}{\partial x} + \frac{\partial \tau_{xy}}{\partial y} = 0 \; ; \qquad \frac{\partial \tau_{yx}}{\partial x} + \frac{\partial \sigma_y}{\partial y} = 0 \, .$$

After integrating along the axis x and y we have

$$\begin{aligned} (\sigma_x)_a &= (\sigma_x)_0 - \int_0^a \frac{\partial \tau_{xy}}{\partial y} dx \approx (\sigma_x)_0 - \sum \frac{\Delta \tau_{xy}}{\Delta y} \Delta x \\ (\sigma_y)_b &= (\sigma_y)_0 - \int_0^b \frac{\partial \tau_{xy}}{\partial x} dy \approx (\sigma_y)_0 - \sum \frac{\Delta \tau_{xy}}{\Delta x} \Delta y \, . \end{aligned} \qquad (1.88)$$

There is one more relation

$$\tau_{xy} = \frac{1}{2}(\sigma_1 - \sigma_2) \sin 2\vartheta \, . \qquad (1.89)$$

Equations (1.88) and (1.89) give the complete solution of the two-dimensional problem on the basis of data obtained experimentally.

1.10.3. Solution by Isopachic Fringe Pattern.

1.10.3.1. Numerical Solution of the Laplace Equation.

The sum of principal stresses $\Sigma = \sigma_1 + \sigma_2$ is governed by the well-known Laplace equation

$$\nabla^2 \Sigma \equiv \frac{\partial^2 \Sigma}{\partial x^2} + \frac{\partial^2 \Sigma}{\partial y^2} = 0 \, . \tag{1.90}$$

By using finite difference method (see Fig.9) the following equation may be written

$$\nabla^2 \Sigma \approx 4\Sigma_0 - (\Sigma_1 + \Sigma_2 + \Sigma_3 + \Sigma_4) \, . \tag{1.90'}$$

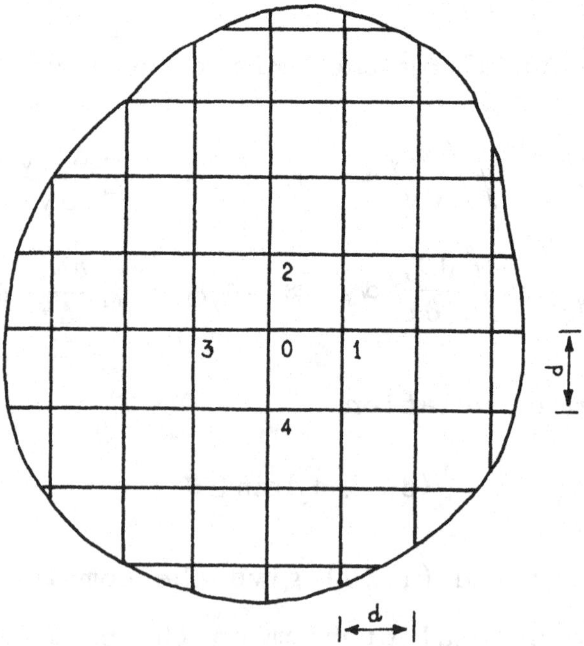

Fig. 9

The solution for the sum $\sum = \sigma_1 + \sigma_2$ at all nodal points of the selected mesh is obtainable by the numerical solution of the corresponding system of algebraic linear equations, and the isopachic lines (the lines of the constant sum of principal stresses) can be traced. Adding the difference $\sigma_1 - \sigma_2$ obtained by using the isochromatic pattern the principal stresses σ_1 and σ_2 can be separated directly.

1.10.3.2. Method of Lateral Extensometer.

The strain in the direction normal to the plane of the considered two-dimensional model is given by the expression

$$\varepsilon_z = -\frac{\nu}{E}(\sigma_1 + \sigma_2)$$

and the change of the thickness d of the model

$$\Delta d = -\frac{\nu}{E}(\sigma_1 + \sigma_2)d \ . \qquad (1.91)$$

By measuring the change Δd (mechanically or optically) one obtains the sum $\sigma_1 + \sigma_2$.

1.10.4. Oblique Incidence Method.

The equation $\sigma_1 - \sigma_2 = i f_\sigma / d$ is based on the light passing through the model at normal incidence. However, if the model is rotated in the polariscope so that the light passes through it at some other angle an oblique incidence fringe pattern may be observed which provides additional data for separating the principal stresses.

If the principal stress directions are known we rotate the model about the σ_1-axis by an amount ϑ, and observe the fringe order i_ϑ due to the secondary principal stresses σ_1 and σ_2'. Thus, (see Fig. 10)

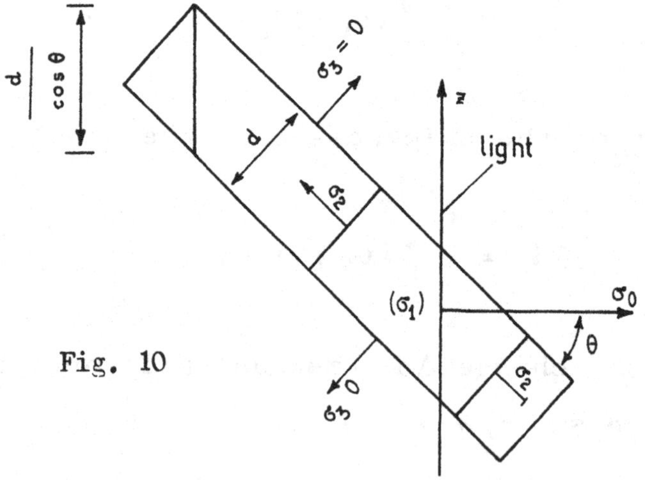

Fig. 10

$$\sigma_1 - \sigma_2' = \frac{f_\sigma i_\vartheta}{d/\cos\vartheta} \tag{1.92}$$

But $\sigma_2' = \sigma_2 \cos^2\vartheta$. By combining last two equations and Eq. (1.64'), we get finally :

$$\sigma_1 = \frac{f_\sigma}{d}\frac{\cos\vartheta}{\sin^2\vartheta}(i_\vartheta - i_0\cos\vartheta)$$
$$\sigma_2 = \frac{f_\sigma}{d}\frac{\cos\vartheta}{\sin^2\vartheta}(i_\vartheta\cos\vartheta - i_0) \qquad (1.93)$$

where i_0 is the fringe order at the normal incidence pattern.

If the directions of principal stresses are not known three isochromatic patterns are necessary : one at normal incidence, and two at oblique incidences under different angles.

1.10.5. Birefringent Coatings.

This method is based upon the bonding of a large, thin plate (sheet) of photoelastic plastic to the surface of specimen. The birefringent coating acts as a strain gage and permits the determination of the principal strain difference over the considered area. The corresponding polariscope is reflective, as shown in Fig. 11.

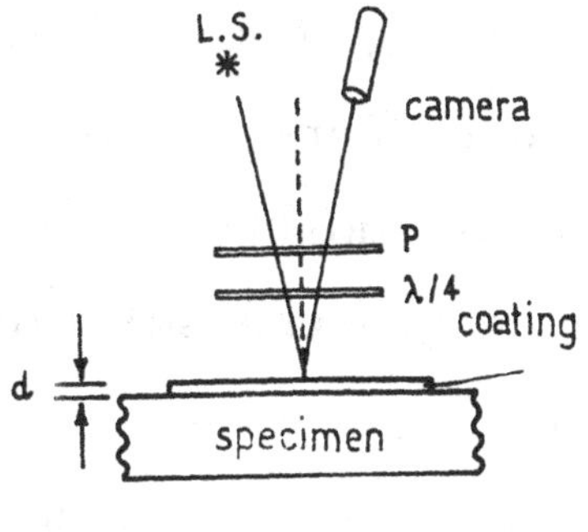

Fig. 11

The basic relation in terms of strain:

$$\varepsilon_1^s - \varepsilon_2^s = \frac{if_\sigma}{2d} = \frac{i}{2d}\frac{1+\nu^c}{E^c}f_\sigma . \tag{1.94}$$

For an elastic body, the conversion to stress formulation is very simple :

$$\sigma_1^s - \sigma_2^s = \frac{E^s}{E^c}\frac{1+\nu^c}{1+\nu^s}\frac{if_\sigma}{2d} \tag{1.95}$$

where the index s means specimen, and c coating.

1.10.6. Interferometric Methods [17].

The interferometry enables the absolute retardation, i.e., the change of refractive indices at each point in a considered photoelastic model and through this, the determination of separated principal stresses throughout the model.

The Favre's absolute retardation method is based on a point-by-point procedure. Recently, Post [17], Nisida-Saito [18], and some other have proposed different types of full-field interferometric methods, mostly in two-dimensional domain, but with chances to the solution of three-dimensional problems too.

In each case of two beam interferometry, optical interference is developed between the active ray (see, f.i., the Mach-Zehnder's interferometer shown schematically in Fig. 12), and the reference ray.

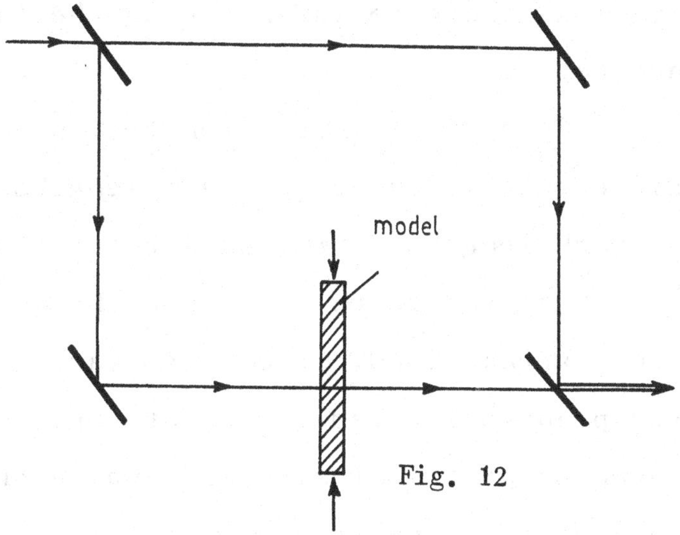

Fig. 12

The resultant interference pattern shows contours along which the difference between the optical-path lengths of the active and reference ray is constant. These rays travel grossly different optical-path lengths, therefore, extremely high coherence, i.e. high monochromatic purity of the illuminating rays, is required to produce interference of suitable optical contrast.

In all the systems using external interferometers, the medium contacting the surfaces of the model may be air with the index of refraction $n = 1$. Alternatively, the medium may be a liquid of arbitrary refractive index or a liquid matching the refractive index of the model in its unstressed state ($n = n_0$). In this latter case, imperfections in the surface of the model material have no influence on the initial interference pattern.

In optical interferometry, fringe order M is the difference in number of wave lengths between the optical-path lengths of the active ray (through the model) and the reference ray. When the model is submitted to external load, it divides each entering ray into two plane-polarized components, polarized in the directions of principal stresses, which propagate through the model with different velocities ; thus, two values of refractive indices (n_{σ_1}, n_{σ_2}) exist for each point of the model. Since optical-path lengths are the products of mechanical thicknesses and refractive indices, there are two optical-path lengths for each point in the birefringent model.

Each active ray splits into two polarized components, which experience different optical-path lengths through the interferometer. The reference ray travels the same optical-path length for all components of polarization and correspondingly polarized components of the active and reference rays come into optical interference, producing interference-fringe orders M_1 and M_2 at each point. Since M_1 and M_2 vary from point to point across the field, the contour maps describing their variations appear simultaneously in the resultant interference pattern.

Using the relations :

(a) $$\Psi_1 = (M_1 - M_0)\lambda \; ; \quad \Psi_2 = (M_2 - M_0)\lambda \qquad (1.96)$$

where Ψ_1 and Ψ_2 are the changes in optical-path lengths induced by the stress system ;

(b) The Maxwell-Neumann stress-optical law :

$$n_1 - n_0 = A\sigma_1 + B\sigma_2 \; ; \quad n_2 - n_0 = B\sigma_1 + A\sigma_2 ; \qquad (1.97)$$

(c) The changes of the optical-path lengths induced by stress alone :

$$\begin{aligned} \Psi_1 &= \beta[n_1 d_s - n_0 d_0 - n(d_s - d_0)] \\ \Psi_2 &= \beta[n_2 d_s - n_0 d_0 - n(d_s - d_0)] \end{aligned} \qquad (1.98)$$

where β is the number of the light passages through the model ;

(d) The change of the thickness (see Fig. 13 page 62) :

$$\varepsilon = \frac{d_s - d_0}{d_0} = -\frac{\nu}{E}(\sigma_1 + \sigma_2) \qquad (1.99)$$

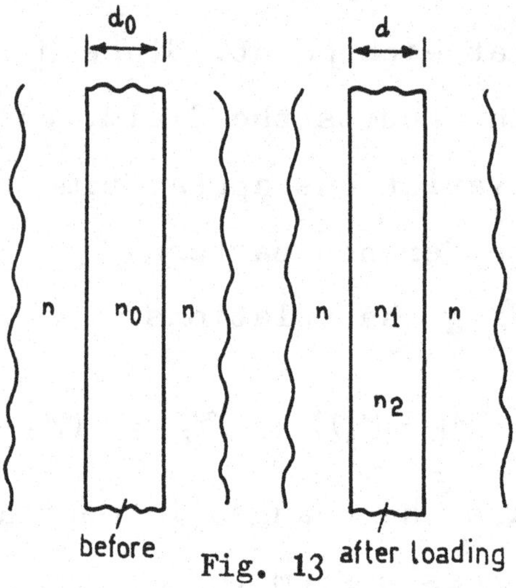

Fig. 13

we are able to deduce the separate expressions for principal stresses :

$$(1.100)\quad \sigma_1 = \frac{1}{d_0}\,\frac{C\Psi_2 - D\Psi_1}{C^2 - D^2}\;; \qquad \sigma_2 = \frac{1}{d_0}\,\frac{C\Psi_1 - D\Psi_2}{C^2 - D^2}$$

where

$$C = \beta\left[A - \frac{\nu}{E}(n_0 - n)\right]; \qquad D = \beta\left[B - \frac{\nu}{E}(n_0 - n)\right]$$

The constants C and D can be obtained directly from a calibration experiment using the same material and apparatus as at the original experiment.

It can be proved that point of intersection of these lines generate contours of constant value $M_1 - M_2$ (Fig. 14 see page 63).

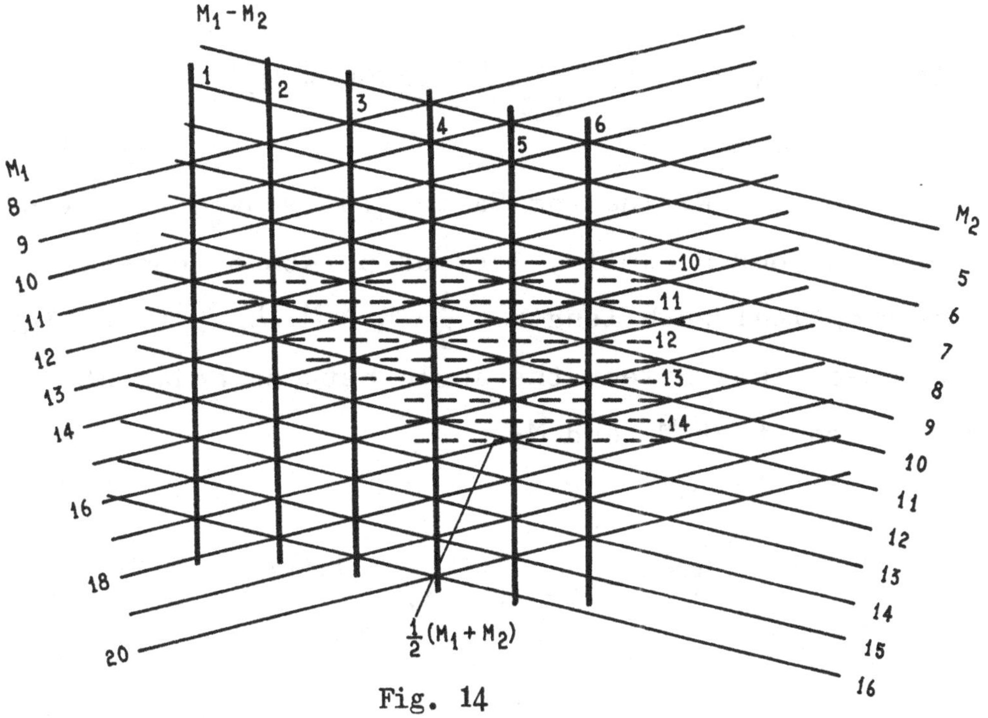

Fig. 14

By Eq. (1.96)

$$(M_1 - M_2)\lambda = \Psi_1 - \Psi_2$$

and by (1.97) and (1.98) is

$$\sigma_1 - \sigma_2 = \frac{\Psi_1 - \Psi_2}{d_0\beta(A - B)} \tag{1.101}$$

The lines $M_1 - M_2$ are identical to the isochromatic lines in photoelasticity :

$$\sigma_1 - \sigma_2 = \frac{\lambda}{d_0\beta(A - B)}(M_1 - M_2) \tag{1.102}$$

Nisida and Saito [18] showed the existence of a fourth family, loci of points of constant average optical-path length $\frac{1}{2}(M_1 + M_2)\lambda$ (see Fig. 15). These lines bisect the angle between adjacent lines of constant M_1 and M_2.

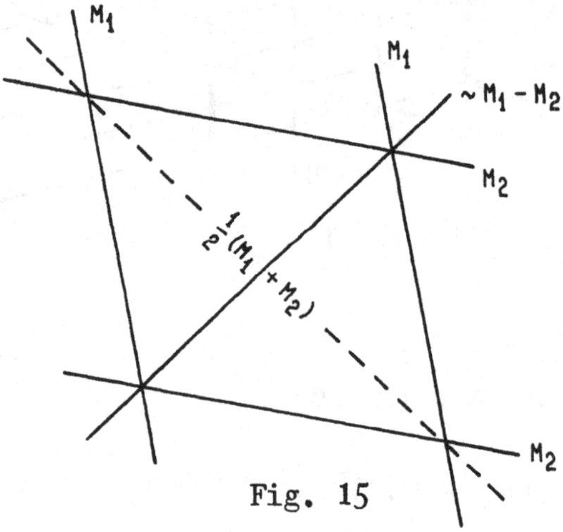

Fig. 15

It can be deduced that

$$\sigma_1 + \sigma_2 = \frac{2}{d_0(C + D)}\frac{\Psi_1 + \Psi_2}{2} \tag{1.103}$$

where the expression $(\Psi_1 + \Psi_2)/2\lambda$ is referred to as the isopachic fringe order. The fringes of constant average optical-path length yield the basic stress parameter

Four parameters of contour lines can be interpreted from the full-load pattern : M_1 , M_2, $M_1 - M_2$, $(M_1 + M_2)/2$. Separate values of M_1 and M_2 can be determined knowing any two of these four parameters, and the separate values of principal stresses σ_1 and σ_2 can be obtained. There is a double redundancy which allows different choices in organizing an experiment.

Using Mach-Zehnder's interferometer and the complex representation of the light amplitude, Nisida and Saito [18] obtained the following result for the light intensity passing through the interferometer :

$$I' = \frac{1}{2} + \frac{1}{2}\cos\left[\frac{2\pi}{\lambda}\frac{A' + B'}{2}(\sigma_1 + \sigma_2)d\right]\cdot \cdot\cos\left[\frac{2\pi}{\lambda}\left\{\frac{C}{2}(\sigma_1 - \sigma_2)\right\}d\right] \qquad (1.104)$$

where

$$A' = A - \frac{\nu}{E}(n_1 - n); \qquad B' = B - \frac{\nu}{E}(n_2 - n);$$
$$C = A - B = A' - B'.$$

The light intensity depends simultaneously as on the value of ($\sigma_1 + \sigma_2$) as on the value of ($\sigma_1 - \sigma_2$) at every point of the considered pattern. That means, the light intensity of isopachic lines is modulated by the intensity of isochromatic lines. After having the values of ($\sigma_1 + \sigma_2$) and ($\sigma_1 - \sigma_2$) the separation of principal stresses is possible directly.

As given by Eq. (1.104) the light intensity at a point where

$$\frac{2\pi}{\lambda} C(\sigma_1 - \sigma_2)d = 2m\pi \qquad (m = \text{integer}) \tag{1.105}$$

i.e. at a point on a line that corresponds to a dark isochromatic line in usual dark-field photoelastic pattern, is expressed as

$$I' = \frac{1}{2} + \frac{1}{2}\cos\frac{2\pi}{\lambda}\frac{A' + B'}{2}(\sigma_1 + \sigma_2)d\ . \tag{1.106}$$

At a point where

$$\frac{2\pi}{\lambda} C(\sigma_1 - \sigma_2)d = (2m + 1)\pi \tag{1.107}$$

i.e. at a point on a line corresponding to a bright isochromatic line in a dark-field photoelastic pattern, it becomes :

$$I' = \frac{1}{2}\ . \tag{1.108}$$

It follows that the half-tone lines that represent the isochromatic lines appear over the pattern running through the family of isopachic fringes.

If the fringe orders of isopachic and isochromatic lines are N_p and N_c, respectively, it is easily seen that the sum and difference of principal stresses can be obtained by

$$\sigma_1 + \sigma_2 = \frac{2\lambda N_p}{d(A' + B')} ; \quad \sigma_1 - \sigma_2 = \frac{2\lambda N_c}{dC} . \quad (1.109)$$

The presented Nisida-Saito's theory is valid for the case of plane stress, i.e. when the stress do not vary along the thickness direction of the model, and when the stress components in that direction are equal to zero.

The constants A' and B' can be easily obtained by using some specimen subjected to simple stress field (simple tension, pure bending, etc.).

1.11. Frozen-Stress Method in Three-Dimensional Photoelasticity. [3]

The general problem of three-dimensional photoelasticity represents one of most tedious tasks in experimental stress analysis. There exists a series of trials to the solution of this problem but the frozen stress method seems to be most complete. The basis of this method is the process of permanently locking the deformations in the model. The deformations are locked in the model on a molecular scale, thus permitting the model to be cut and sliced without relieving the locked-in deformations. The three-dimensional problem is deduced that way into a series of two-dimensional problems by cutting the considered model into thin plane slices which are analyzed then as plane, two-dimensional models.

The theoretical basis of the frozen stress method is the diphase behaviour of many polymeric materials when they are heated. Some of the molecular chains of these materials are well bonded into a three-dimensional network of primary bonds (see Fig. 16). However, a large number of molecules are less solidly bonded together into shorter secondary chains.

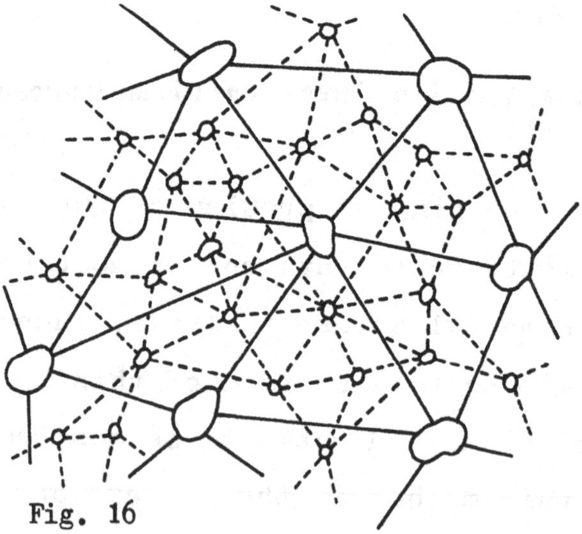

Fig. 16

When the polymer is at room temperature, both sets of molecular bonds, the primary and the secondary, act to resist deformation due to external load. However, as the temperature of the polymer is increased, the secondary bonds break down and the primary bonds in effect carry the entire applied load. Since the secondary bonds constitute a very large portion of the polymer, the deflections which the primary bonds undergo are quite large yet elastic in character. If the temperature of the polymer is lowered to room temperature while the load is maintained on the model, the secondary bonds will reform between the highly elongated primary bonds and serve to lock them into their extend positions. When the load is removed, the primary bonds relax to a modest degree, but the main portion of their deformation is not recovered. The elastic deformation of the primary bonds is permanently locked into the model by the reformed secondary bonds. Moreover, these deformations and accompanying birefringence are maintained in any small section cut from the original model.

At this stage the photoelastic model with its locked-in deformations and attendant fringe pattern may be carefully cut or sliced without disturbing the character of either the deformation or the fringe pattern.

The temperature to which the model is heated to break down the secondary bonds is called the critical temperature.

The three-dimensional model with locked-in deformations is then sliced to remove the planes of interest which can then be examined individually to determine the state of stress existing in that particular plane or slice (which must be sufficiently thin in relation to the size of the model that the stress do not change in either magnitude or direction through the thickness of the slice).

In the first order, the slices on free boundary or in the planes of symmetry can be examined directly as two-dimensional model, since one has the principal stresses and not the secondary principal stresses here,and the problem is two-dimensional one. At some general location, the problem has to be treated as three-dimensional, and three slices at a point are necessary in order to obtain all stress components at the considered point.

Chapter 2.
Photoviscoelasticity.

2. 1. Introduction.

The basic mathematical theory of the problem of photoviscoelasticity was given by R. D. Mindlin [20] (1949) in his well-known paper "A Mathematical Theory of Photo-Viscoelasticity". He considered the time dependent relations between three second rank tensors : birefringence, stress, and strain. His emphasis was mainly upon the conditions under which a viscoelastic model material could be used to deduce the elastic stress distribution in a prototype. Upon assuming elastic volume change he found that, if the material was incompressible and the loading was proportional (i.e. no variation in the distribution pattern of the load but only a magnitude change with the time), that in the relations

$$n = C_\sigma(\sigma_1 - \sigma_2) = 2C_\sigma \tau_{max}$$

$$n = C_\varepsilon(\varepsilon_1 - \varepsilon_2) = C_\varepsilon \gamma_{max}$$

the stress-optical and the strain-optical constants are viewed as time dependent operators characteristic of the material. The polarizing axes and the principal axes were aligned in this case.

A year later, W. T. Read [21] published a paper concerned to the generalized viscoelastic analysis of compressible media. He demonstrated that, as in Photoelasticity, the optical measurements give the directions and difference in magnitude of the principal stresses in the plane of the optical wavefront, provided the stress-optical relations can be expressed in an operational form as in the stress-strain relations. The Read's formulation placed no restrictions on the boundary conditions. The principal axes are not necessarily aligned with the polarizing axes.

The analysis of propagation of viscoelastic waves (Arenz, see Ref. [22]) showed that the glassy and rubbery states correspond to elastic conditions, so that optical and mechanical principal axes are known to be aligned at the start and finish of the wave propagation process. Hence, axis nonalignment will occur at most during the photoviscoelastic transition period. Therefore, it can be assumed, for engineering purposes that the mechanical and optical axes are always aligned, and the experimental disposition should be organized in a manner for which all axes are aligned a priori.

Mindlin and Read treated the problem of photoviscoelasticity from the standpoint of Mechanics, macroscopically. But many problems connected with the molecular structures of polymers are concerned to the question upon the degree to which optical, and also mechanical anisotropy on the microscale affects the gross macroproperties of concern to the stress analyst. This microaspect of the problem, taking into account the anisotropy in both birefringence and strain, is an important objective of chemical investigations.

For analysis, it is necessary to have the optical and mechanical characterization of viscoelastic materials as a function of strain rate and temperature. The stiffness of viscoelastic materials is dependent essentially upon the temperature of the material and the rate at which the load is imposed.

In the paper published by M. L. Williams and R. J. Arenz [23], where the general engineering analysis of photoviscoelastic materials has been given, it was assumed that birefringent-material response may be considered as linearly viscoelastic for engineering purposes.

Further progress in the analysis of photoviscoelastic materials was made by Theocaris and

Mylonas [24] who pointed out a quantitative but empirical relation between strain rate and temperature which reflects the fact that behaviour at high temperatures and high strain rates is similar to that at low temperatures and low strain rates. This fact, called sometimes time-temperature superposition principle, is the basis of the possibility of time reduction.

E.H. Dill [25] gave the analysis of Photoviscoelasticity and Photothermoviscoelasticity using the approach of Continuum Mechanics and proposing general possibility of the experimental verification of results.

A very exhaustive engineering analysis of photoviscoelastic materials is given in the beforementioned paper by Williams and Arenz [23] with the accentuation to the experimental aspect of the problem, especially in connection with the application to dynamical problems.

Many more papers being published recently (I. Daniel [26,27], J.T. Pindera [28], V. Brčić and M. Nešović [29],[30] , M.G. Sharma and C.K. Lim [31], R.M. Hackett and E.M. Krokosky [32], H.F. Brinson [33]) are concerned to the technology of materials being applied to different fields of photoviscoelasticity in both static and dynamic cases, and to the solution of practical engineering problems.

2.2. The Mindlin's and the Read's Approach to Photoviscoelasticity.

2.2.1. The Mindlin's Theory of Photoviscoelasticity.

R.D. Mindlin [20] started with the supposition of linearity, losing the possibility of analyzing finite strain and nonlinear stress-strain and stress-strain-rate relations and without introducing thermodynamic effects. By using the general strain-optical relation for a viscoelastic model with m elastic elements and a number of viscous elements, assuming that the birefringence arises from the deformation of elastic elements only, he started with the relation

$$K_{ij} - n_0 \delta_{ij} = \sum_{s=1}^{m} (c_s e_{ij} + \frac{1}{3} d_s \varepsilon_{kk} \delta_{ij}) \qquad (2.1)$$

where

K_{ij} is the tensor of dielectric constants,

$e_{ij} = \varepsilon_{ij} - \frac{1}{3} \delta_{ij} \varepsilon_{kk}$ is the deviatoric part of the strain tensor, c_s , d_s are strain-optical coefficients, n_0 is index of refraction of the unstrained medium.

The object of the Mindlin's analysis was a four-elemental model (see Fig. 17 page 76),

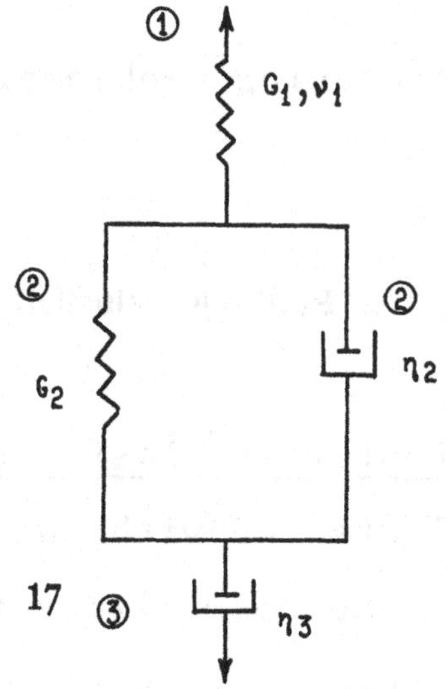

Fig. 17

where the part 1 represents instantaneous elasticity, the part 2 retarded elasticity, and the part 3 flow. The assumption of incompressibility was applied. The constitutive equation of such an element has the form :

$$(2.2) \qquad P(s_{ij}) = 2Q(e_{ij})$$

where

$$s_{ij} = \sigma_{ij} - \frac{1}{3}\delta_{ij}\sigma_{kk}$$

is the deviatoric part of the stress tensor, and P ( ) and Q ( ) are known differential operators depending on constants G_1 , G_2 , ν_1 , η_2 , η_3 .

In the case of such idealized four-elemental viscoelastic medium Eq. (2.1) becomes

$$(2.3) \qquad K_{ij} - n_0^{-2}\delta_{ij} = c_1(e_{ij})_1 + c_2(e_{ij})_2 + \frac{1}{3}d_1 e\,\delta_{ij} .$$

The <u>three</u> strain-optical coefficients c_1 , c_2 and d_1 are necessary to describe the optical properties of this viscoelastic medium. Two of them, c_1 and c_2 , are <u>relative</u> strain-optical coefficients and contribute only to relative phase retardation. The third coefficient, d_1, is a mean strain-optical coefficient whose magnitude can be determined only by measurement of absolute phase retardation. All of the coefficients may be temperature dependent.

By combining Eqs. (2.3.) and (2.2) Mindlin obtained the stress-optical equation in the form

$$K_{ij} - n_0^{-2}\delta_{ij} = R(s_{ij}) + 2S(e_{ij}) + \frac{1}{3}d_1\delta_{ij}\varepsilon_{kk} \qquad (2.4)$$

where R and S are differential operators (see Ref.20).

In the case of a medium exhibiting flow, the relative birefringence depends upon the stress, stress-rate, and strain-rate, and is independent of the strain itself. In the cases of simpler models (f.i: without flow, or without retarded elasticity) the dependence on some of these quantities can be absent.

The optical quantities which have to be measured are relative phase retardation ϱ and amplitude ratio γ.

In the case of homogeneous birefringent body, the relative phase retardation is given by

$$\varrho = 2\pi L \frac{n_1 - n_2}{\lambda} \qquad (2.5)$$

where L is the path length, λ is the wave length in vacuum, and n_1 and n_2 are the principal refractive indices in the plane of the wave-front. The amplitude ratio gives the angle φ which a polarizing axis makes with a reference direction in the plane of the wave-front.

In the case when the stress and strain deviators can be expressed in the form of product

$$(2.6) \qquad \begin{aligned} s_{ij} &= \bar{s}_{ij}(x,y,z)\, f(t) \\ e_{ij} &= \bar{e}_{ij}(x,y,z)\, g(t) \end{aligned}$$

one obtains the relation

$$(2.7) \qquad n_1 - n_2 = \overline{(\sigma_1 - \sigma_2)}\, C(t)$$

where

$$(2.8) \qquad C(t) = \frac{1}{2} n_0^3 \left\{ R(f) + \frac{S(g)}{G_0} \right\} .$$

G_0 is an arbitrary constant of the nature and dimensions of a shear modulus.

The relation $s_{ij} = \bar{s}_{ij} f(t)$ leads to a simple stress-optical relation for viscoelastic materials completely analogous to that for an elastic medium. The time and temperature-dependent relative stress-optical coefficient $C(t)$ can be determined by a specimen subjected to simple tension where the tensible stress σ is applied according to the desired law $\bar{\sigma} f(t)$ and the temperature is varied with time in a desired manner.

Similarly the strain-optical relation:

$$n_1 - n_2 = \overline{(\varepsilon_1 - \varepsilon_2)}\, c(t) \tag{2.9}$$

where

$$c(t) = n_0^3 \{G_0 R(f) + S(g)\} . \tag{2.10}$$

$c(t)$ is obtainable similarly, by an elongation experiment in which the strain ε follows a prescribed law $\bar{\varepsilon}\, g(t)$, and the temperature is varied with time in the desired manner.

Due to Mindlin, the following conditions are necessary in order to, at a viscoelastic body, for the stress deviator to be of the form $\bar{s}_{ij} f(t)$ and for $\bar{s}_{ij}$ to correspond to the stress in an elastic body of the same shape under analogous boundary conditions :

(1) The viscoelatic material must be incompressible.

(2) The boundary conditions must be of the specific form as in elasticity theory (some specific combinations of forces and displacements).

(3) Components of the boundary forces, when prescribed, must have the form

$$\bar{T}_j(x,y,z)f(t)$$

and components of boundary displacements

$$\bar{U}_j(x,y,z)\,g(t)$$

where $Q(g) = G_0 P(f)$

(4) The initial stress must be equal to zero.

Under these conditions the stress and strain in viscoelastic body are related to the birefringence by Eqs. (2.7) and (2.9), and the axes of stress, strain and birefringence are aligned. The barred components in (2.7) and (2.9) of stress and strain are those for an incompressible elastic body with shear modulus G_0.

2.2.2. The Read's Theory of Compressible Viscoelastic Media.

W.T. Read [21] proposed the theory for linear, compressible viscoelastic (or anelastic) materials. For such materials stress, strain, and their time derivatives of all orders are related by linear equations with coefficients which are material constants. Read used the Fourier integral method showing that static elasticity solutions can be to determine the time dependent stresses in viscoelastic bodies with <u>any</u> form of boundary conditions.

The most general linear isotropic stress-strain relations can be expressed in terms of three linear differential operators H, K and L :

$$H\left(\frac{\partial}{\partial t}\right)\sigma_x = K\left(\frac{\partial}{\partial t}\right)e + 2L\left(\frac{\partial}{\partial t}\right)\varepsilon_x, \ldots, \ldots, \qquad (a)$$

(2.11)

$$H\left(\frac{\partial}{\partial t}\right)\tau_{xy} = L\left(\frac{\partial}{\partial t}\right)\gamma_{xy}, \ldots, \ldots, \qquad (b) .$$

The operators have the form :

$$H\left(\frac{\partial}{\partial t}\right) = H_0 + H_1\left(\frac{\partial}{\partial t}\right) + H_2\left(\frac{\partial^2}{\partial t^2}\right) + \ldots + H_n\left(\frac{\partial^n}{\partial t^n}\right) + \ldots$$

H_0, H_1, ... are constants of the material, obtained by using the model with the corresponding number of springs and dashpots.

In the general case a more practical approach is to determine two complex constants $G(i\omega)$ and $\lambda(i\omega)$ which are functions of frequency ω and may be determined from a simple experiment and used to express the stress-strain relations as follows.

We represent the time dependence of

$\tau_{xy} = \tau_{xy}(t)$ by a Fourier integral

(2.12) $$\tau_{xy}(t) = \int_{-\infty}^{+\infty} \bar{\tau}_{xy}(\omega) e^{i\omega t} d\omega$$

where

(2.12) $$\bar{\tau}_{xy}(\omega) = \frac{1}{2\tau} \int_{-\infty}^{+\infty} \tau_{xy}(t) e^{-i\omega t} dt$$

is a Fourier transform of $\tau_{xy}(t)$ Expressing γ also as a Fourier integral, we have

(2.13) $$\bar{\tau}_{xy}(\omega) = G(i\omega) \bar{\gamma}_{xy}(\omega)$$

where the complex modulus

(2.14) $$G(i\omega) = \frac{L\left(\frac{\partial}{\partial t}\right) e^{i\omega t}}{H\left(\frac{\partial}{\partial t}\right) e^{i\omega t}} = \frac{L(i\omega)}{H(i\omega)} .$$

Defining another complex modulus

(2.15) $$\lambda(i\omega) = \frac{K(i\omega)}{H(i\omega)}$$

the relation (2.11a) becomes :

(2.16) $$\bar{\sigma}_x(\omega) = \lambda(i\omega)\bar{e}(\omega) + 2G(i\omega)\bar{e}_x(\omega) .$$

Thus, the time dependent stress-strain relations (2.11) are replaced by static relations between the Fourier transforms of stress and strain. Of course, the complex Young's modulus $E(i\omega)$ and Poisson's ratio $\nu(i\omega)$, which are related to $\lambda(i\omega)$ and $G(i\omega)$ by the same way as in elastic case, can be used too. The determination of those complex constants can be performed by a simple tension test in which the tension $\sigma_x(t)$ varies with time in any convenient manner. By measuring the strain $e_x(t)$ and $e_y(t)$ (parallel and perpendicular to tension), and taking the Fourier transforms of the three known functions of time, we obtain :

$$E(i\omega) = \frac{\bar{\sigma}_x(\omega)}{\bar{e}_x(\omega)} ; \qquad \nu(i\omega) = -\frac{\bar{e}_y(\omega)}{\bar{e}_x(\omega)} . \qquad (2.17)$$

The same can be generalized for anisotropic materials too.

By replacing elastic constants λ and G by complex constants $\lambda(i\omega)$ and $G(i\omega)$ we may the solution of a viscoelastic problem reduce to the solution of the equivalent elastic problem, supposing that the boundary conditions have also the correspondent time variation.

If the body forces can be neglected and the boundary conditions vary uniformly with time, and if the time dependence of the applied forces and displacements is $F(t)$, then the solution for a general viscoelastic solid is

$$(2.18) \qquad S(x,y,z,t) = \int_{-\infty}^{+\infty} \bar{F}(\omega) S_e\left[x,y,z,\lambda(i\omega),G(i\omega)\right] e^{i\omega t}\, d\omega$$

where

$$(2.18) \qquad \bar{F}(\omega) = \int_{-\infty}^{+\infty} F(t) e^{-i\omega t}\, dt$$

and $S_e(x, y, z, \lambda, G)$ is the elastic solution for $F(t) = 1$. $\bar{S}$ may be any components of stress, strain or displacement.

When the elasticity solution involves only the ratio of elastic moduli, conveniently expressed by the dimensionless quantity ν , special methods and simple approximate procedures can be used. It can be taken that a general linear solid is equivalent to an elastic solid with a time variable Poisson's ratio.

Another simple case occurs in problems of plane strain where the boundary conditions are given in stresses and there is no resultant force on internal boundary ; the elasticity solution is then independent of the elastic constants and the stress distribution in general case is $F(t)$ times the static elasticity solution.

Read had studied the optical relations at viscoelastic bodies for the case that the stress-optical relations are of the same form as the stress-strain relations, and can be expressed by linear differential equations with constant coefficients. It is shown in this case that the standard photo-elastic methods can be employed to determine the directions of the secondary principal axes and the secondary principal stress difference in the plane of the wave-front. No restrictions on the boundary conditions are involved.

Let V_{ij} be the velocity tensor. If the xy-plane is the plane of the wave-front, then the polarizing axes are the axes of the ellipse formed by the trace of the ellipsoid in the xy-plane, and the principal wave velocities are inversely proportional to the squares of the major and minor axes of the ellipse. The standard photoelasticity gives the difference $V_1 - V_2$ and the angle φ between the polarizing axes and the fixed x-axis. The difference $V_{xx} - V_{yy}$ is given by the relation

$$V_{xx} - V_{yy} = (V_1 - V_2)\cos 2\varphi \qquad (2.19)$$

where all quantities vary with both time and position.

For an isotropic viscoelastic material the most general linear differential equations for the time dependence of the stress-optical effect give

$$X(\sigma_x - \sigma_y) = Y(V_{xx} - V_{yy}) \tag{2.20}$$

where X and Y are linear differential operators which may have any number of terms. The last two equations give

$$\sigma_x - \sigma_y = \frac{Y}{X}(V_1 - V_2)\cos 2\varphi \tag{2.21}$$

from which the stress difference $\sigma_x - \sigma_y$ can be calculated as a function of time for all points where $(V_1 - V_2)$ and φ have been measured.

As in photoelasticity, the optical measurements give the directions and difference in magnitude of the secondary principal stresses. Generally, the principal axes of stresses are not parallel to the polarizing axes at viscoelastic bodies.

Especially, in the case where the state of stresses is known to be the same as the stress in an elastic body, Eq. (2.21) takes the simple form :

$$\sigma_1 - \sigma_2 = K(t)(V_1 - V_2) \tag{2.22}$$

where

$$K(t) = \frac{1}{F(t)} \frac{Y}{X} F(t)$$

and σ_1 and σ_2 are the secondary principal stresses which act in the directions of the polarizing axes. By a simple tension test the function $K(t)$ can be obtained.

2. 3. General Theory of Photoviscoelasticity. [25]

2.3.1. Fundamental Equations of Viscoelasticity.

The principle of conservation of mass leads to the field equation

$$\frac{\partial \varrho}{\partial t} + div(\varrho \underset{\sim}{v}) = 0 \qquad (2.23)$$

where ϱ is the density and $\underset{\sim}{v}$ is the velocity vector.

The stress vector $\underset{\sim}{\sigma}$ on a surface with unit normal $\underset{\sim}{n}$ is related to the stress tensor σ^{km} by the equation

$$\sigma^k_{(\underset{\sim}{n})} = \sigma^{km} n_m \, . \qquad (2.24)$$

The principle of conservation of linear and angular momentum leads to the equation of motion

$$(2.25) \qquad \sigma^{km}_{,m} + \rho f^{k} = \rho a^{k}$$
$$\sigma^{km} = \sigma^{mk}$$

where f is the body force, a is the acceleration.

Generally, the electromagnetic field gives rise to the body force

$$\underset{\sim}{f} = Q\underset{\sim}{E} + \underset{\sim}{I} \times \underset{\sim}{B}$$

where

Q is the charge density

$\underset{\sim}{I}$ is the current density

$\underset{\sim}{B}$ is the density of magnetic flux.

<u>Constitutive equations for deformation</u> :

The deviatoric stress s_{ij} is defined by

$$s_{ij} = \sigma_{ij} - \frac{1}{3}\sigma_{kk}\,\delta_{ij}$$

where $p = -\frac{1}{3}\sigma_{kk}$ is the mean pressure.

The deviatoric strain :

$$e_{ij} = \varepsilon_{ij} - \frac{1}{3}\varepsilon_{kk}\,\delta_{ij}$$

For small strains, the volume dilatation is

$$\frac{d\vartheta - dV}{dV} \approx \varepsilon_{kk} = e \, .$$

By empirical evidence, it follows that, for some materials experiencing isothermal deformations and small displacements, the material is characterized with sufficient accuracy by the relations :

$$e_{ij} = \int_{-\infty}^{t} I(t - x) \frac{ds_{ij}(x)}{dx} dx$$
$$e = - \int_{-\infty}^{t} \varkappa(t - x) \frac{dp(x)}{dx} dx \qquad (2.26)$$

or, in the inverse form :

$$s_{ij} = \int_{-\infty}^{t} G(t - x) \frac{de_{ij}(x)}{dx} dx$$
$$p = - \int_{-\infty}^{t} K(t - x) \frac{de(x)}{dx} dx \qquad (2.27)$$

where the kernels of these Duhamel integrals are :

$G(t)$ relaxation modulus in shear

$I(t)$ creep compliance in shear

$K(t)$ bulk relaxation modulus

$\varkappa(t)$ bulk creep compliance .

Sometimes, the following alternative

notation is more convenient :

$$G(t) = G_0 + \Phi(t)$$

$$I(t) = I_0 + \frac{t}{\eta} + \psi(t)$$

where

I_0 is the initial elastic compliance

η is a Newtonian viscosity coefficient

Ψ is the creep function $\Psi(0) = 0$

G_0 is the equilibrium elastic modulus

ϕ is the relaxation function, $\phi(\infty) = 0$.

This notation has the advantage that the creep function is generally found to be a *monotonically* increasing function with initial value zero and having monotonically decreasing slope. The relaxation function is then a *monotonically* decreasing function tending to zero and having monotonically increasing (algebraically) slope.

Each of these quantities is temperature dependent as well as time dependent so they should be written in the form $G(t,T)$ where T is the absolute temperature. For *isothermal* deformations, empirical evidence suggests that, for many materials the temperature dependence can be adequately described with the aid of a "temperature shift function".Namely, it is found that, for test at any given temperature, if a reduced time τ defined by

$$\ln\tau = \ln t - f(t) \tag{2.28}$$

is introduced, then

$$G\{t(\tau)\} = \bar{G}(\tau) \tag{2.29}$$

where $\bar{G}(t)$ is the relaxation modulus at a reference temperature T_0. The function $f(T)$ is a monotonically decreasing function of temperature such that $f(T_0) = 0$ and is regarded as a material property. The shift factor $a(t)$ is defined by

$$\ln\alpha = -f(T)$$

i.e.

$$\tau = \alpha(T)t\ . \tag{2.30}$$

Similar relations are assumed to hold for bulk modulus. A material characterized by Eq. (2.29) is said to be thermorheologically simple.

Eq. (2.29) implies that

$$G(0) = \bar{G}(0) = G_0 + \Phi(0)$$

and

$$G(\infty) = \bar{G}(\infty) = G_0\ .$$

Thus, the initial and final elastic moduli are not affected by temperature change. Thus, only the relaxation function has to be considered :

$$(2.31) \qquad \phi(t) = \phi(\tau)$$

where $\bar{\phi}(t)$ is the relaxation function at the reference temperature T_0.

Similarly, for creep compliance :

$$(2.32) \qquad I(t) = \bar{I}(\tau)$$

implies I_0 and η are independent of temperature and

$$(2.33) \qquad \Psi(t) = \bar{\Psi}(\tau)$$

where $\bar{\Psi}(t)$ is the creep function at the reference temperature T_0.

2.3.2. Constitutive Equations for Linear Thermoviscoelasticity.

There is a viscous dissipation of energy but if the heat due to this dissipation can be neglected and since the volume changes are small, the temperature distribution is determined by the uncoupled Fourier heat conduction equation :

$$(2.34) \qquad T_{,kk} = K\frac{\partial T}{\partial t} .$$

Thus, the temperature may be regarded as a known function of space and time, $T = T(x,t)$. The theory is then rendered complete by satisfying a phenomenological relation between stress, strain and temperature.

The shear creep compliance measured by a creep test at varying temperature is $f(t)$. If the initial and final compliance for a material not experiencing viscous flow are independent of temperature, then $f(0) = \bar{\Gamma}(0)$ and $f(\infty) = \bar{\Gamma}(\infty)$. In this case there is a function $\xi(t)$ called the reduced time such that

$$f(t) = \bar{\Gamma}\{\xi(t)\} \qquad (2.35)$$

and

$$f'(t) = \frac{d\xi}{dt}\,\bar{\Gamma}'(\xi)$$

where the prime indicates the differentiation with respect to the argument.

The <u>isothermal</u> creep test of a thermorheologically simple material, Eq. (2.32), yields :

$$\Gamma'(t) = a(T)\,\bar{\Gamma}'(\tau) . \qquad (2.36)$$

It is physically appealing to assume that this relation holds also for the creep at <u>non-constant temperature</u>.

With this fundamental hypothesis

$$\frac{d\xi}{dt} = a\{T(t)\}$$

Therefore

$$\xi(t) = \int_0^t a\{T(x)\}\,dx \;. \tag{2.37}$$

Now suppose that the same temperature field exists, but the load is applied at time τ. The resulting creep compliance is $F(t,\tau)$ The total strain resulting from a continuous distribution of load is then

$$e_{ij}(t) = \int_0^t F(t,x)\frac{ds_{ij}(x)}{dx}\,dx \;. \tag{2.38}$$

With assumption $F(\tau,\tau) = \bar{I}(0)$, and $F(\infty,\tau) = \bar{I}(\infty)$, there exists a function $\xi(t,\tau)$ such that

$$F(t,\tau) = \bar{I}(\xi) \tag{2.39}$$

and

$$\frac{dF(t,\tau)}{dt} = \frac{d\xi}{dt}\,\bar{I}'(\xi) \;.$$

According to fundamental hypothesis

$$\frac{d\xi}{dt} = a\{T(t)\} \;.$$

Since $F(\tau,\tau) = \bar{I}(0)$, $\xi = 0$ at $t = \tau$. Therefore :

$$\xi(t,\tau) = \int_{\tau}^{t} a\{T(\eta)\}\,d\eta = \xi(t) - \xi(\tau) \qquad (2.40)$$

and

$$F(t,\tau) = \bar{I}\{\xi(t) - \xi(\tau)\} . \qquad (2.41)$$

Introducing the notation

$$\bar{e}_{ij}\{\xi(t)\} = e_{ij}(t) ; \qquad \bar{s}_{ij}\{\xi(t)\} = s_{ij}(t) \qquad (2.42)$$

the constitutive equation becomes :

$$\bar{e}_{ij}(\xi) = \int_{0}^{\xi} \bar{I}(\xi - \eta)\,\frac{d\bar{s}_{ij}(\eta)}{d\eta}\,d\eta . \qquad (2.43)$$

The fundamental constitutive equation for thermoviscoelasticity is therefore Eq. (2.38) which under the fundamental hypothesis can be put into the form (2.43) for thermorheologically simple materials.

2. 4. Photothermoviscoelasticity. [25]

2.4.1. Dielectric Constitutive Equations for Isothermal Deformation.

Experiments show that certain polymers

exhibit polarization under isothermal deformations which depend upon the history of deformation. This suggests that a relationship exists between the dielectric tensor and the strain tensor similar to that for linear viscoelasticity.

Let N_{ij} be the refraction tensor whose principal axes coincide with those of the dielectric tensor K_{ij}, and such that the principal values N_i of N_{ij} are related to the principal values K_i of K_{ij} by

$$N_i = \sqrt{K_i} .$$

Denote the deviatoric part of N_{ij} and let

$$N = \frac{1}{3} N_{kk}$$

Consider the hypothesis

$$\bar{N}_{ij} = \int_0^t \chi(t - \tau) \frac{d}{d\tau}\{e_{ij}(\tau)\} d\tau \tag{2.44}$$

$$N = \int_0^t \Delta(t - \tau) \frac{d}{d\tau}\{e(\tau)\} d\tau + N_0 \tag{2.44a}$$

where

$\chi(t)$ is the optical-relaxation function

$\Delta(t)$ is the optical-volume-relaxation function.

Denote the Laplace transform of a function by a superposed asterisk.

Then :

$$\bar{N}^*_{ij} = p\chi^* e^*_{ij} \tag{2.45}$$

where p is the transform variable. From Eq. (2.26):

$$e^*_{ij} = pI^* s^*_{ij} \tag{2.46}$$

Therefore

$$s^*_{ij} = p\Gamma^* \bar{N}^*_{ij} \tag{2.47}$$

where

$$\Gamma^* = \frac{1}{p^3\chi^* I^*}$$

The inverse transform gives :

$$s_{ij}(t) = \int_0^t \Gamma(t - \tau)\frac{d}{d\tau}\{\bar{N}_{ij}(\tau)\} d\tau \ . \tag{2.48}$$

Thus, a unique relation between the polarization and the history of deformation implies a unique relation between the stress and <u>the history of polarization</u> and inversely.

Both the function $\chi(t)$ and $\Gamma(t)$ may be expected to depend on the temperature. Analogous to the situation for the mechanical state, existence of an <u>optical-temperature shift factor</u> $b(T)$

may be supposed such that

$$\chi(t) = \bar{\chi}\{b(T)t\} \tag{2.49}$$

where $\bar{\chi}(t)$ is the optical-relaxation function at temperature T_0, and $\chi_0(t)$ is the optical-relaxation function at temperature T. Similarly :

$$\Gamma(t) = \bar{\Gamma}\{b(T)t\} \tag{2.50}$$

2.4.2. Dielectric Constitutive Equations for Thermal Stress.

Consider deformations which occur in the presence of varying temperatures $T(t)$. If the polarization at various strain levels but the same temperature history can be superimposed, the constitutive equation for the isothermal deformations (2.44) becomes

$$\bar{N}_{ij} = \int_0^t f(t,\tau)\frac{d}{d\tau}\{e_{ij}(\tau)\}d\tau . \tag{2.51}$$

The function $f(t,\tau)$ may in general depend on the history of temperature.

In the case of thermorheologically simple materials, it may be that a reduced time $\varrho(t)$ exists, in general different from that for mechanical case, such that

$$f(t,\tau) = \bar{X}\{\varrho(t,\tau)\} \tag{2.52}$$

$$\varrho(t,\tau) = \int_{\tau}^{t} b\{T(t)\}\,dt\ . \tag{2.53}$$

In this event, the optical properties are completely characterized by optical-relaxation function at reference temperature T_0, $\bar{X}(t)$, and the optical-temperature shift factor $b(T)$. Equation (2.51) can be written in a form similar to Eq. (2.41) :

$$\bar{\bar{N}}_{ij}(\eta) = \int_{0}^{\eta} \bar{X}(\eta - \tau)\frac{d}{d\tau}\{\bar{\bar{e}}_{ij}(\tau)\}\,d\tau \tag{2.54}$$

where

$$\eta(t) = \int_{0}^{t} b\{T(t)\}\,dt$$

$$\bar{\bar{N}}_{ij}(\eta) = \bar{N}_{ij}\{t(\eta)\}$$

$$\bar{\bar{e}}_{ij}(\eta) = e_{ij}\{t(\eta)\}$$

If the temperature shift factors $a(T)$ and $b(T)$ are different, no simple transformation to the form analogous to Eq. (2.48) exists. But if $a(T) = b(T)$, then $\eta(t) = b(t)$. Taking the Laplace transform with respect to ξ of Eq.(2.54) and Eq. (2.43) gives:

(2.55) $$\bar{\bar{N}}^*_{ij} = p\bar{X}^*\bar{e}^*_{ij}$$

(2.56) $$\bar{e}^*_{ij} = p\bar{I}^*\bar{s}^*_{ij} .$$

Therefore

(2.57) $$s^*_{ij} = p\bar{\Gamma}^*\bar{\bar{N}}^*_{ij}$$

where

$$\bar{\Gamma}^* = \frac{1}{p^3\bar{X}^*\bar{I}^*} .$$

The inverse transformation gives :

(2.58) $$\bar{s}_{ij}(\xi) = \int_0^{\xi} \bar{\Gamma}(\xi - \tau)\frac{d}{d\tau}\{\bar{\bar{N}}_{ij}(\tau)\}\,d\tau .$$

2.4.3. Principal Axes.

Let z_i be the fixed rectangular coordinate axes with $z_3 = z$. Consider the plane stress problem and take $z_3 = z$ normal to the plane of the stress field. Suppose that z is a principal axis of σ_{ij} , ε_{ij} and N_{ij}. Let x_1 be the principal axis of strain, $x_3 = z_3$, and let α be the angle from z_1 to x_1. Let y_1 be the principal axis of stress, $y_3 = z_3$, and let β be the angle from z_1 to y_1. Let X_1 be the principal axis of N_{ij}, $X_3 = z_3$, and let γ be the angle from z_1 to X_1 (Fig. 18).

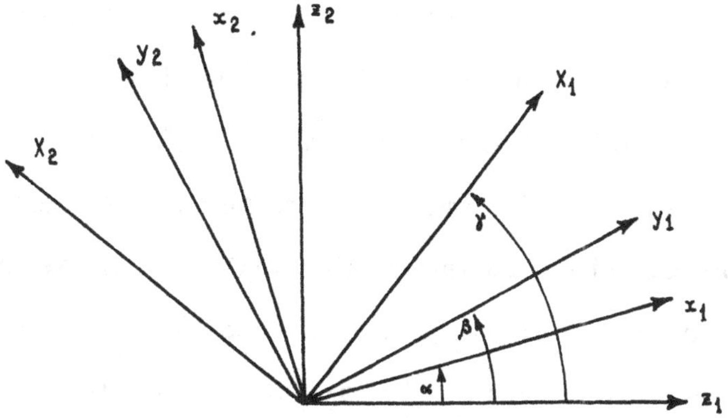

Fig. 18

It follows from Eq. (2.48) that

$$\sigma_{11}(t) - \sigma_{22}(t) = \int_0^t \Gamma(t-\tau)\frac{d}{d\tau}\Big\{\big[N_1(\tau) - N_2(\tau)\big]\cos 2\gamma(\tau)\Big\}d\tau \qquad (2.59)$$

$$\sigma_{12}(t) = \int_0^t \Gamma(t-\tau)\frac{d}{d\tau}\Big\{\big[N_1(\tau) - N_2(\tau)\big]\sin 2\gamma(\tau)\Big\}d\tau\ . \qquad (2.60)$$

It is seen that a knowledge of the angle $\gamma(t)$ and the difference $N_1(t) - N_2(t)$, of the principal value of N_{ij}, as a function of time is necessary and sufficient to determine the difference in principal stresses and the principal axes of stress in the isothermal case.

A similar reasoning applies to the transient case when Eq. (2.58) may be used.

2.4.4. Experimental Procedure.

The model is placed in a standard plane polariscope and subjected to a desired loading and heating sequence. The isoclinic and isochromatic patterns will vary with time. A complete record of the time variation of the fringe order $n(t)$ and the axis of polarization $\gamma(t)$ at a point is made. This will require, of course, a more elaborate recording device than is normally used with the polariscope. For a given model thickness h, Eqs. (1.41) and (1.36a),

$$n \approx \frac{h\omega}{2\pi c}(N_2 - N_1) \; ; \quad \tan\delta_n = \frac{1 + N_n^2}{2N_n}\tan k_n h$$

determine a value for $N_1(t) - N_2(t)$. Eqs. (2.59) and (2.60) may then be used in the isothermal case to determine the difference in principal stresses $\sigma_1(t) - \sigma_2(t)$ and the axis of principal stresses $\beta(t)$ as function of time. To determine the material properties, this procedure is applied to any configuration for which the stress is known, f.i. a creep test.

2. 5. The Engineering Analysis of Linear Photoviscoelastic Materials.[23]

2.5.1. Mechanical Properties of Photoviscoelastic Materials [34].

2.5.1.1. Introductory Remarks. The mechanical Properties of an isotropic, homogeneous perfectly elastic body are characterized by two material constants (any two of E , ν , G , K , λ, μ). Let be taken the bulk modulus K and the shear modulus G as two material constants.

The deviatoric stress and strain components are given by the expressions :

$$s_{ij} = \sigma_{ij} - \frac{1}{3}\delta_{ij}\sigma_{kk} ; \quad e_{ij} = \varepsilon_{ij} - \frac{1}{3}\delta_{ij}\varepsilon_{kk} .$$

In this form, the distortion, or shear, relation becomes :

$$s_{ij} = 2Ge_{ij} \qquad (2.61a)$$

and the dilatation, or bulk relation

$$\sigma_{ii} = 3K\varepsilon_{ii} . \qquad (2.61b)$$

The linear viscoelastic medium is characterized by an isothermal stress-strain relation given in the form of a linear-differential operator, associated with a corresponding rheological model (with springs and dashpots) :

$$(2.62)\quad \begin{aligned} \left(a_n \frac{\partial^n}{\partial t^n} + \cdots + a_s\right) s_{ij}(x_k,t) &= \left(b_m \frac{\partial^m}{\partial t^m} + \cdots + b_0\right) e_{ij}(x_k,t) \\ \left(c_r \frac{\partial^r}{\partial t^r} + \cdots + c_0\right) \sigma_{ii}(x_k,t) &= \left(d_s \frac{\partial^s}{\partial t^s} + \cdots + d_0\right) \varepsilon_{ii}(x_k,t) \end{aligned}$$

where the material constants a_i , b_i , c_i , d_i have to be determined experimentally, f.i. by specifying the stress response to specified bulk and shear inputs.

Applying Laplace's transformation of Eq. (2.62) we get

$$(2.63)\quad \begin{aligned} \bar{s}_{ij}(x_k,p) &= \frac{b_m p^m + \cdots + b_0}{a_n p^n + \cdots + a_0} \bar{e}_{ij}(x_k,p) \equiv 2G(p)\bar{e}_{ij}(x_k,p) \\ \bar{\sigma}_{ii}(x_k,p) &= \frac{d_s p^s + \cdots + d_0}{c_r p^r + \cdots + c_0} \bar{\varepsilon}_{ii}(x_k,p) \equiv 3K(p)\bar{\varepsilon}_{ii}(x_k,p) \end{aligned}$$

or inversely :

$$(2.64)\quad \begin{aligned} &\text{(a)} \qquad 2\bar{e}_{ij}(x_k,p) \equiv I(p)\bar{s}_{ij}(x_k,p) \\ &\text{(b)} \qquad 3\bar{\varepsilon}_{ii}(x_k,p) \equiv B(p)\bar{\sigma}_{ii}(x_k,p) \end{aligned}$$

which reveal pseudo-elastic character of the stress-strain law in the transform plane by comparison with Eqs. (2.61a) and (2.61b). The purpose of material characterization is to determine the nature of $G(p)$

and $K(p)$ in a convenient form which can be used in stress analysis. Instead of $G(p)$ and $K(p)$ the following material constants may be used too :

$$E(p) = \frac{9G(p)K(p)}{3K(p) + G(p)} \; ; \quad \nu(p) = \frac{3K(p) - 2G(p)}{6K(p) + 2G(p)} \; . \qquad (2.65)$$

It may be assumed that rubbery materials are incompressible (i.e. $\nu = 1/2$ for small strains). Because of limited experimental possibilities it is convenient to consider a three stages of approximation to the material behaviour. The first is incompressibility in bulk but permitting viscoelastic shear behaviour. The second permits a finite value of the bulk modulus but neglects any time dependence (i.e. replacing is actual time-dependence by an average constant (elastic) behaviour. The shear behaviour is assumed viscoelastic as before. The last stage is to assume that both bulk and shear are viscoelastic, but linear, isotropic and homogeneous.

The first stage gives :

$$K(p) = \infty \; ; \quad \nu(p) = \frac{1}{2} \; ; \quad E(p) = 3G(p) \; . \qquad (2.66)$$

The second stage gives :

$$K(p) = K_e \; ; \quad \nu(p) = \frac{3K_e - 2G(p)}{6K_e + 2G(p)} \; ; \quad E(p) = \frac{9K_e G(p)}{3K_e + G(p)} \; . \qquad (2.67)$$

The incorporation of either assumption usually depends upon the availability of experimental data.

2.5.1.2. Methods of Characterization.

There are four possible tests which can be used in order to characterize the material properties :

(a) Step-strain input : measure stress-relaxation output (Fig. 19)

(b) Step-stress input : measure creep-strain output (Fig. 20).

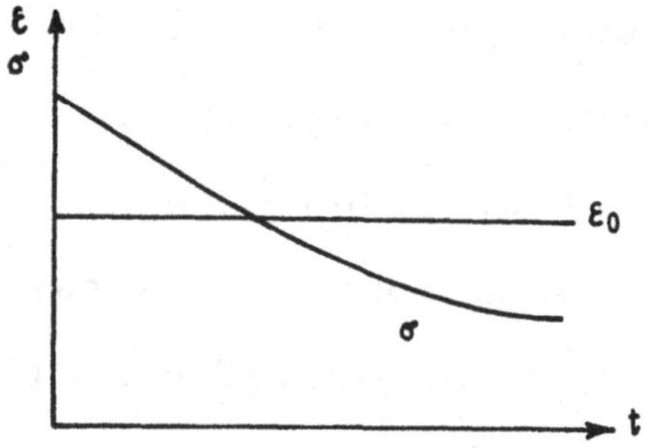

Fig. 19

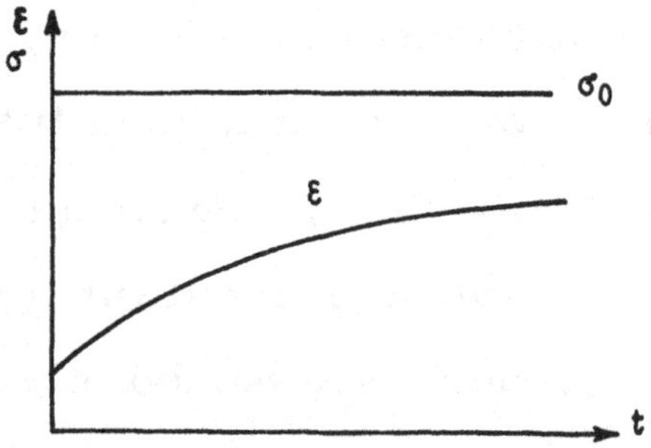

Fig. 20

(c) Constant strain-rate input : measure stress output (Fig. 21).

(d) Sinusoidal stress (strain) input : measure strain (stress) output (Fig. 22), (see page 107).

The specimen can be a beam in simple tension or a narrow cantilever beam. The practical achievement of strict initial conditions (f.i. step-strain) requires a very precise laboratory arrangement.

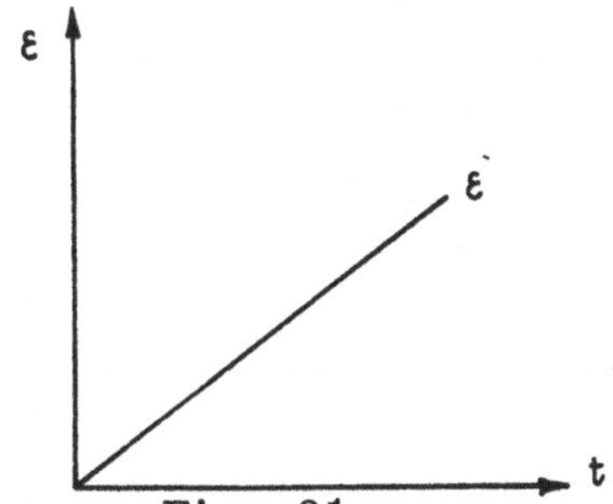

Fig. 21

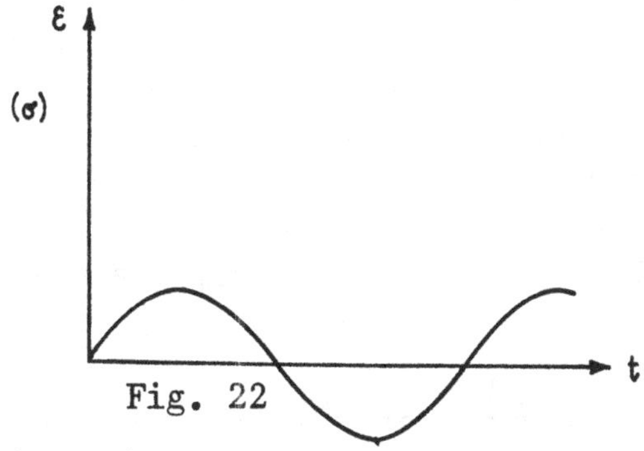

Fig. 22

Write Eq. (2.63) in terms of principal stresses and strain :

$$\frac{1}{2}\left[\bar{\sigma}_1(p) - \bar{\sigma}_2(p)\right] = G(p)\left[\bar{\varepsilon}_1(p) - \bar{\varepsilon}_2(p)\right] \qquad (2.68)$$

and consider a simple uniaxial-tension test of an incompressible material. Thus,

$$\begin{aligned} &\sigma_1 = \sigma; \ \sigma_2 = 0; \ \varepsilon_1 = \varepsilon; \ \varepsilon_2 = -\nu\varepsilon_1; \ \nu = 1/2 \\ &\bar{\sigma}(p) = 3G(p)\bar{\varepsilon}(p) = E(p)\bar{\varepsilon}(p) \end{aligned} \qquad (2.69)$$

This relation will be used in further examples ; it is the viscoelastic analog of the uniaxial elastic tension test.

(a) Stress Relaxation.

Assume $\varepsilon(t) = \varepsilon_0$ (for $t > 0$) ; then $\bar{\varepsilon}(p) = \varepsilon_0/p$. Upon defining the relaxation modulus

$$E_{rel}(t) = \frac{\sigma_{rel}(t)}{E_0}; \quad \bar{E}_{rel}(p) = \frac{\bar{\sigma}_{rel}(p)}{\varepsilon_0} \qquad (2.70)$$

as the normalized stress relaxation to the step strain, we have using (2.69) :

$$\bar{\sigma}_{rel}(p) = E(p)\left[\frac{\varepsilon_0}{p}\right]$$

giving

$$E(p) = p\bar{E}_{rel}(p) . \tag{2.71}$$

(b) Creep.

By using Eq. (2.64a) and noting the invers of $E(p)$ is defined as the compliance $D(p)$, one finds that, due to a step-stress input $\sigma(t) = \sigma_0$, then $\bar{\sigma}(p) = \sigma_0/p$. Upon defining the creep compliance

$$D_{crp}(t) = \frac{1}{\sigma_0}\varepsilon_{crp}(t)\,;\quad D_{crp}(p) = \frac{1}{\sigma_0}\bar{\varepsilon}_{crp}(p) \tag{2.72}$$

as the normalized creep response to an applied step stress, one obtains, upon inverting Eq. (2.69) :

$$\bar{\varepsilon}(p) = D(p)\bar{\sigma}(p) \tag{2.73}$$

or

$$D(p) = pD_{crp}(p) . \tag{2.74}$$

(c) Constant Strain Rate.

This form is relatively easy to be obtained in laboratory by using a testing machine imposing a constant head speed. One requires :

$$\varepsilon(t) = Rt \; ; \quad \bar{\varepsilon}(p) = \frac{R}{p^2} \; .$$

The stress response is $\sigma_{ten}(t)$ such that Eq. (2.69) gives :

$$\frac{1}{R} p\bar{\sigma}_{ten}(p) = \frac{1}{p} E(p) = \bar{E}_{rel}(p) \qquad \text{(using 2.71)} \; . \quad (2.75)$$

But the left term is the transform of the slope of the tensile stress-strain curve during the constant strain-rate test starting from rest :

$$L\left[\frac{d\sigma_{ten}(t)}{RdT}\right] = \frac{p\bar{\sigma}_{ten}(p)}{R} \qquad (2.76)$$

which is precisely the relaxation modulus,

$$\left.\frac{d\sigma_{ten}(t)}{d\varepsilon}\right|_{\varepsilon = Rt} = E_{rel}(t) \qquad (2.77)$$

where $\varepsilon = Rt$ is the tensile strain during the constant-strain test.

The statistical theory of rubber-like elasticity shows that the internal elastic refractive forces in polymeric chains are directly proportional to the absolute temperature T. Hence, the test results can be normalized to an arbitrary reference temperature T_0, if the stress in Eq. (2.77) is adjusted by the ratio T_0/T. Then, upon introducing R in both numerator and denominator and using logarithmic derivatives, Eq. (2.77) can be written to include isothermal test conditions as

$$\frac{T_0}{T}E_{rel}(t) = \left\{\frac{\sigma T_0/RT}{\varepsilon/R}\,\frac{d[\log(\sigma T_0/RT)]}{d[\log(\varepsilon/R)]}\right\}_{t=\varepsilon/R} . \tag{2.78}$$

(d) Dynamic responses.

An oscillating stress or strain input at constant frequency is to be applied. Returning to the linear differential operator from Eq. (2.69) as in Eq. (2.62) we have :

$$\left[a_n\frac{\partial^n}{\partial t^n} + \cdots + a_0\right]\sigma(t) = 3\left[b_m\frac{\partial^m}{\partial t^m} + \cdots + b_0\right]\varepsilon(t) . \tag{2.79}$$

There are two possibilities : (1) the strain is specified as $\varepsilon_0 e^{i\omega t}$ where ε_0 is real number representing the maximum amplitude of the sine wave, and the response is $\sigma^* e^{i\omega t}$ where σ^* is a complex function of frequency, or (2) oppositely, the stress prescribed as $\sigma_0 e^{i\omega t}$ and strain as $\varepsilon^* e^{i\omega t}$.

Upon substituting in (2.79) for the

first case, we get :

$$\left[a_n(i\omega)^n + \cdots + a_0\right]\sigma^* e^{i\omega t} = 3\left[b_m(i\omega)^m + \cdots + b_0\right]\varepsilon_0 e^{i\omega t}$$

so that, upon introducing the definition of complex modulus :

$$\frac{\sigma^*(\omega)}{\varepsilon_0} \equiv E^*(\omega) = E'(\omega) + iE''(\omega) = \frac{3\left[b_m(i\omega)^m + \cdots + b_0\right]}{\left[a_n(i\omega)^n + \cdots + a_0\right]} \quad (2.80)$$

Similarly, the complex compliance to the stress input is conveniently defined as

$$\frac{1}{\sigma_0}\varepsilon^*(\omega) \equiv D^*(\omega) = D'(\omega) - iD''(\omega) = \left[\frac{1}{E^*(\omega)}\right]. \quad (2.81)$$

2.5.1.3. Mathematical Model of Linear Viscoelastic Medium.

The simplest model possessing visco-elastic properties is the stress-element model, known as Poynting-Thomson's model, as represented in Fig. 23. It exhibits an instantaneous glassy elasticity as well as creep and recovery. There are only three material constants. From Fig. 23 (see page 112) it is seen that

$$\varepsilon = \varepsilon_s + \varepsilon_\alpha ; \quad \sigma = \sigma_e + \sigma_m$$

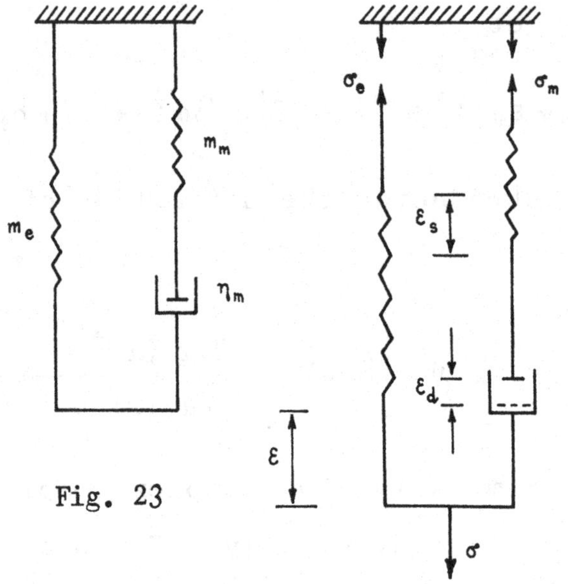

Fig. 23

where

$$\frac{d\varepsilon_s}{dt} = \frac{1}{m_m}\frac{d\sigma_m}{dt} ; \quad \frac{d\varepsilon_d}{dt} = \frac{\sigma_m}{\eta_m} ; \quad \varepsilon = \frac{\sigma_e}{m_e} .$$

The operator equation has for this model the form :

$$\sigma(t) = \left\{ m_e + \frac{m_m \frac{d}{dt}}{\left(\frac{d}{dt} + \frac{1}{\tau_m}\right)} \right\} \varepsilon(t)$$

or

$$\sigma(t) = \frac{m_g\left(\frac{d}{dt} + \frac{m_e}{m_g}\frac{1}{\tau_m}\right)}{\left(\frac{d}{dt} + \frac{1}{\tau_m}\right)} \varepsilon(t)$$

where

$$\tau_m = \frac{\eta_m}{m_m} ; \qquad m_g = m_m + m_e .$$

Creep and Recovery (Fig. 24)

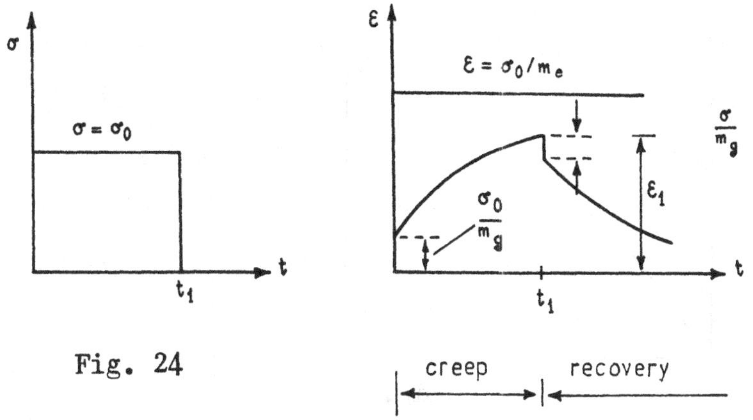

Fig. 24

$$\varepsilon = \left\{1 - \left(1 - \frac{m_e}{m_g}\right) e^{-\frac{m_e}{m_g}\frac{t}{\tau_m}}\right\} \frac{\sigma_0}{m_e} \qquad (0 < t \leq t_1)$$

$$\varepsilon = \left\{\varepsilon_1 - \frac{\sigma_0}{m_g}\right\} e^{-\frac{m_e}{m_g}\frac{t - t_1}{\tau_m}} \qquad (t > t_1)$$

Relaxation (Fig. 25)

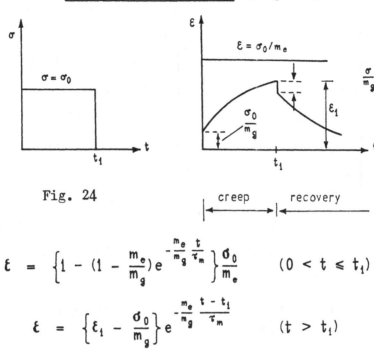

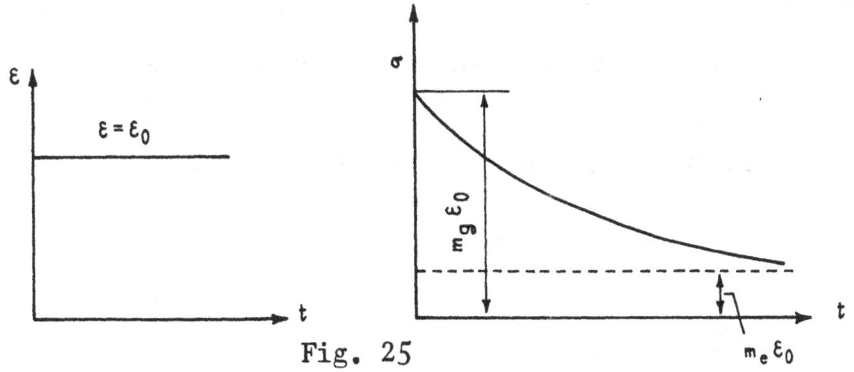

Fig. 25

$$\sigma = \left\{1 + \left(\frac{m_g}{m_e} - 1\right) e^{-\frac{t}{\tau_m}}\right\} m_e \varepsilon_0 \qquad (t > 0)$$

Constant Strain Rate (Fig.26)

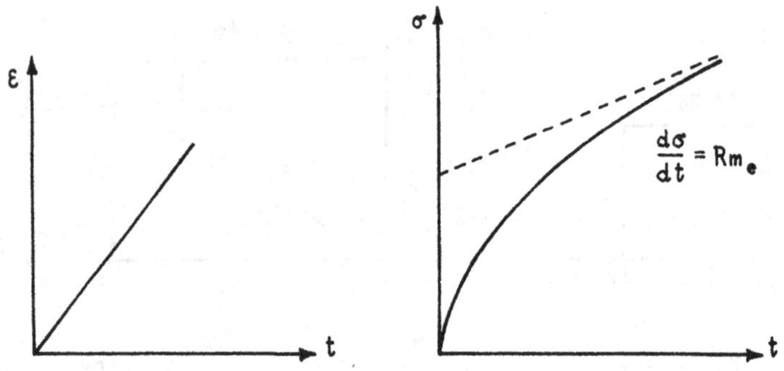

Fig. 26

$\varepsilon = Rt \quad (R = \text{strain rate})$

$$\sigma = \left\{\frac{t}{\tau_m} + \left(\frac{m_g}{m_e} - 1\right)\left(1 - e^{-\frac{t}{\tau_m}}\right)\right\} R m_e \tau_m \; .$$

It is also possible to compose the model having an arbitrary number of spring-dashpot elements (Wiechert's model, see Fig. 27), or some different combinations of springs and dashpots (in series or combined).

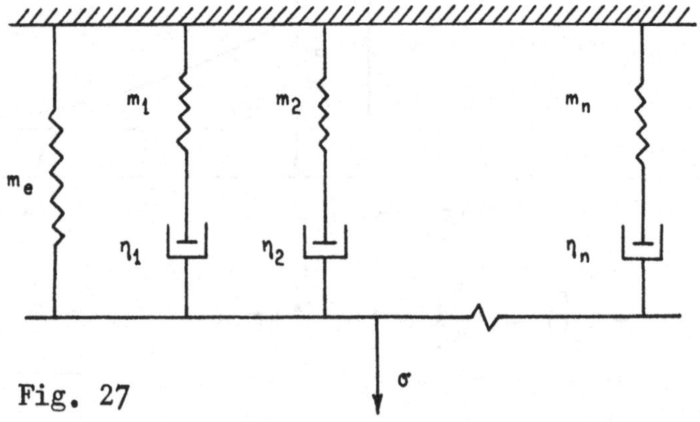

Fig. 27

While it is true that the accuracy in representation over a larger range of time interval is increased, one must remember that one of the must important purposes of data representation is to use the final mathematical formulas in stress analysis. For example, $E(p) = 3G(p)$ is the ratio of polynomials which require subsequent Laplace inversions, and the computations can become rather tedious. It is desirable therefore to have some methods of representation in both the real an transformed variable which is not only accurate but tractable. For example, the Laplace transform of the relaxation modulus $E_{rel}(t)$ is defined

$$\bar{E}_{rel}(p) = \int_0^\infty E_{rel}(t) e^{-pt} dt .$$

It follows that

$$E(p) = p\bar{E}_{rel}(p) .$$

R.A. Shapery [35] has taken advantage of the decaying exponential character of viscoelastic deformations to propose an effective collocation scheme. Choosing the relaxation modulus for example, he assumes that

$$E_{rel}(t) = E_e + \sum E_k e^{-\frac{t}{\tau_k}} \qquad (2.82)$$

which Dirichlet series expansion can be thought of as an arbitrary expansion of the function $E_{rel}(t)$ into a "Fourier" series of real exponentials having the Fourier coefficients E_k and modulus (i/τ_k) instead of the usual integer k. Under certain conditions, this expansion is mathematically complete, although it is not possible to write down explicitly the solution for the coefficients as in usual Fourier analysis. Shapery selects N decades of time over the transition range of the experimental data and sets the range of τ such that one τ_k falls in each decade over this range. The experimental data are collocated at N points , which leads to a simply solved triangular matrix. The Laplace transform of Eq.(2.82) is straight forward

$$(2.83) \qquad E(p) = p\bar{E}_{rel}(p) = E_e + \sum \frac{p\tau_k E_k}{p\tau_k + 1}$$

and, essentially, permits the factorization of the ratio of the two polynomials, Eq. (2.63), in an expeditious manner.

It may be noted that the spectrum of retardation times is distributed over the transition range at the discrete, real value τ_k . If $p = i\omega$ the complex modulus can also be computed.

It can be shown that $D_{rel}(t)$ can be obtained from the curve fit of $E_{rel}(t)$.

It was mentioned already that the empirical results show a characteristical property of polymers, the behaviour at low temperatures and low strain rates is similar to that at high temperatures and high strain rates. Therefore, the temperature-shift function can be introduced with the aim to deal in terms of a reduced or reference time which is independent of temperature. The usual form of this relation, as proposed by Williams, Landel and Ferry (WLF-formula), is :

$$\log a_T(T) \equiv \log\left(\frac{t}{t_R}\right) = -\frac{K_1 \frac{T}{T_R}}{K_2 + (T - T_R)} \qquad (2.84)$$

where t is the physical time to observe some phenomenon at the temperature $T(^\circ K)$ compared to the time for a similar phenomenon at a reference temperature T_R. Materials which obey this law are called, as mentioned earlier, thermorheologically simple.

For many polymers it has been found experimentally that $K_1 = 8{,}86$ and $K_2 = 101{,}6$, and the reference temperature T_R is approximately $50\,^\circ C$ above the glass temperature of the material.

By multiplying numerator and denominator by the constant rate R, i. e.

$$a_T = \frac{Rt}{Rt_R} = \frac{\varepsilon_T}{\varepsilon_0}$$

it can be shown that under constant strain-rate conditions Eq. (2.84) can be interpreted as relating strains at two different temperatures using the reduced time parameter $t' = t/a_T(T)$..

For instance, the shift factor can then be incorporated into the relaxation modulus as measured in the constant-strain-rate test, accounting for the similarity of low temperature-low strain rate and high temperature-high strain rate behaviour. Similar to Eq. (2.78), it results in the following equation in terms of reduced time, defined as $t' = \varepsilon/(Ra_{T_0})$, and reduced strain rate Ra_{T_0}:

$$(2.85)\quad \frac{T_0}{T} E_{rel}(t') = \left\{ \frac{\sigma T_0/(Ra_{T_0}T)}{\varepsilon/(Ra_{T_0})} \frac{d[\log(\sigma T_0/Ra_{T_0}T)]}{d[\log(\varepsilon/Ra_{T_0})]} \right\}_{t'=\varepsilon/Ra_{T_0}}$$

This reduced-time parameter t' can be used in the analysis of photoviscoelastic data.

The term shift factor arises from the way in which a_T is determined experimentally. Theocaris and Mylonas [38] investigated the behaviour of a cold-setting epoxy resin and tested its stress relaxation due to constant strain input measured at various temperatures over the usual four or five decades of time and plotted on semilog paper as represented in Fig. 28a. If one held the upper cold-temperature curve fixed and translated the lower curves to the right by an increasing amount depending upon the temperature, a continuous smooth curve("master curve") would result from the "temperature shift".

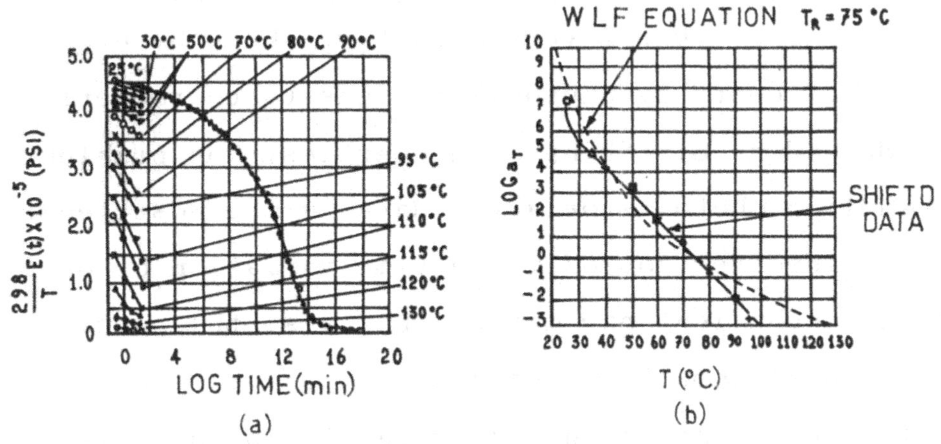

Fig. 28

One can then plot a graph of the decades of temperature shift from the reference low temperature as a function of temperature (Fig. 28b). This graph, prepared from the data of Fig. 28a in the form of $\log a_T$ vs. T and adjusted arbitrary for zero shift at

75 °C (50 °C above ambient temperature) is called the shift curve. It is frequently well approximated by the WLF expression

$$\log a_T(T) \equiv \log\left(\frac{t}{t_R}\right) = -\frac{K_1 \frac{T}{T_R}}{K_2 + (T - T_R)}$$

except at the higher temperatures (see Fig. 28b).

2.5.2. Analysis of Optical Properties of Photoviscoelastic materials.

Until very recently, there have been no much attempts by photoelasticiens to characterize time-dependent birefringent properties. The problem of handling photoviscoelastic data in the most general case involves not only the time variation of optical and mechanical parameters, but also the possibility of nonalignment of the axes of polarization, principal stresses and principal strains. In the case of isotropic, homogeneous, viscoelastic media, in which the coefficients become time operators, special conditions must apply before the alignment of all pertinent axes is assured. For viscoelastic materials where the stress-and strain-optical relations are expressed by the type of linear operators discussed previously for mechanical properties, the directions of principal stress axes and difference in principal stresses can be determined, as discussed, f.i. in the Read's paper [21]. It requires the measurement of both isochromatics ans isoclinics as functions of time. Maximization of the resulting $(\sigma_x - \sigma_y)$ values with respect to the orientation of the arbitrary coordinate axes gives the difference in principal stresses and their directions.

Therefore, the time varying isoclinics and isochromatics become a necessary part of the experimental data. The basic knowledge of the optical-operator relationship between birefringence and stress or strain is required. This is the basic characterization of the material behaviour.

In order to obtain it a corresponding experimental arrangement, f.i. simple-tension test, is necessary. Here, the alignment of optical and mechanical axes is assured beforehand. Thus, the birefringe relationships are simplified and can be formulated directly as functions of the difference in principal strains or stresses.

As discussed earlier (see. Ch. 2.2.2.) the standard photoelastic techniques can be used to determine the directions and the difference of principal stresses.

We have the relation

$$(2.86)\quad \sigma_x - \sigma_y = \frac{Y}{X}(V_1 - V_2)\cos 2\varphi = \frac{\beta_m \frac{\partial^m}{\partial t^m} + \cdots + \beta_0}{\alpha'_n \frac{\partial^n}{\partial t^n} + \cdots + \alpha'_0}(V_1 - V_2)\cos 2\varphi$$

and taking into account the relation between principal velocities and relative retardation, n, and the operator ratio as C_σ^{-1} we have :

$$(2.87)\qquad \sigma_x - \sigma_y = C_\sigma^{-1}[n\cos 2\varphi]$$

from which $(\sigma_x - \sigma_y)$ can be calculated providing the optical-property characterization C_σ^{-1} is known and the relative retardation and isoclinic variation have been recorded as function of time.

As shown by Dill [25], one more equation can be written by resolving the stress and velocities in terms of shear as

$$(2.88)\qquad 2\tau_{xy} = C_\sigma^{-1}[n\sin 2\varphi] .$$

Using the well-known relations

$$(2.89)\qquad \sigma_1 - \sigma_2 = \frac{\sigma_x - \sigma_y}{\cos 2\Psi} ; \quad \tan 2\Psi = \frac{2\tau_{xy}}{\sigma_x - \sigma_y}$$

where Ψ is the angle between σ_1 and x-axis, we have the equations in operator form :

$$(\sigma_1 - \sigma_2)\cos 2\Psi = C_\sigma^{-1}[n\cos 2\Psi]$$
$$(\sigma_1 - \sigma_2)\sin 2\Psi = C_\sigma^{-1}[n\sin 2\Psi] \qquad (2.90)$$

from which $(\sigma_1 - \sigma_2)$ and Ψ can be obtained.

In the general case of a photovisco-elastic medium, the angles φ and Ψ will not coincide, but when they do, the last equation gives :

$$\sigma_1 - \sigma_2 = C_\sigma^{-1}[n] \qquad (2.91)$$

or, after applying Laplace transforms,

$$\bar{n}(x_i,p) = C_\sigma(p)[\bar{\sigma}_1(x_i,p) - \bar{\sigma}_2(x_i,p)] \ . \qquad (2.92)$$

Assuming again aligned axes, similar relation can be written for strain :

$$\varepsilon_1 - \varepsilon_2 = C_\varepsilon^{-1}[n] \qquad (2.93)$$

or

$$\bar{n}(x_i,p) = C_\varepsilon(p)[\bar{\varepsilon}_1(x_i,p) - \bar{\varepsilon}_2(x_i,p)] \qquad (2.94)$$

2.5.2.1. Determination of Relaxation-Birefringence-Strain Coefficient.

Proceeding therefore with the case of the aligned optical and mechanical axes, i.e. $\varphi = \Psi$, for all time, consider the Eq. (2.94) relating relative retardation with principal strain difference. Note, that C_ε is related to the usual photoelastic material strain-optic coefficient f_ε, based on maximum shear strain, as

$$C_\varepsilon = \frac{h}{2f_\varepsilon}$$

where h is the model thickness.

In a <u>relaxation test</u>, f.i., the corresponding birefringence-strain coefficient can be defined as the fringe response due to a constant-strain-difference input,

$$C_{\varepsilon_{rel}}(t) = \frac{n_{rel}(t)}{(\varepsilon_1 - \varepsilon_2)_0} \quad (2.95)$$

However, it is difficult to impose a perfect step-strain-difference input, therefore, a constant-strain-rate test is proposed. The two results are connected as in Eq. (2.77) and an essentially similar derivation for the relaxation birefringence-strain coefficient gives :

$$C_{\varepsilon_{rel}}(t) = \frac{2}{3}\frac{1}{R}\frac{dn_{tens}}{dt} \quad (2.96)$$

or, upon incorporating the time-temperature shift factor in terms of reduced time,

$$C_{\varepsilon_{rel}}(t') = \frac{2}{3}\left\{\frac{n/(Ra_{T_0})}{\varepsilon/(Ra_{T_0})}\frac{d\,\log(n/Ra_{T_0})}{d\,\log(\varepsilon/Ra_{T_0})}\right\}_{k' = \varepsilon/Ra_{T_0}} \tag{2.97}$$

Here, the material is assumed as incompressible ($\nu = 1/2$), and the test in uniaxial, so that if $\varepsilon = \varepsilon_1$ is the axial strain, and ε_2 the transverse, then

$$\varepsilon_1 - \varepsilon_2 = \varepsilon_1 - (-\nu\varepsilon_1) = \frac{3}{2}\varepsilon_1 .$$

Values of $\log(n/R)$ vs. $\log(\varepsilon/R)$ for each isothermal run are to be plotted and then shifted according to standard procedure, to form a smooth master curve of $\log(a/Ra_{T_0})$ vs. $\log t'$. The time-temperature shift factor a_T thus obtained can be plotted and compared to the shift factor curve determined for the mechanical data.

2.5.2.2. Conversion to Creep Birefringence-Stress Coefficient.

A biregringence-stress coefficient can be defined as in Eq. (2.91) and is related to the usual material stress-optic coefficient f_σ by $C_\sigma = h/(2f_\sigma)$. The creep birefringence-stress differential operator, $C_{\sigma_{crp}}(t)$, which would be determined during a constant stress-difference test, is not obtainable from a constant-strain-rate test. But, it can be obtained from $C_{\varepsilon_{rel}}(t)$ for linear viscoelastic material since interrelationships among the time-dependent functions in Eqs. (2.92) and (2.94) can be derived.

For illustration, let the time-dependent shear response be charaterized by

$$P(\sigma_1 - \sigma_2) = Q(\varepsilon_1 - \varepsilon_2) \tag{2.98}$$

where P and Q are linear differential operators. If reduced time is used in Eq. (2.98), then the Laplace transforms of Eq. (2.92) and (2.94) are related. Upon combining Eqs. (2.93) and (2.98) we have

$$\bar{n}(p) = C_\sigma(p)[\bar{\sigma}_1(p) - \bar{\sigma}_2(p)] \tag{2.99}$$

or:

$$\bar{n}(p) = C_\sigma(p)\frac{Q(p)}{P(p)}[\bar{\varepsilon}_1(p) - \bar{\varepsilon}_2(p)] = C_\varepsilon(p)[\bar{\varepsilon}_1(p) - \bar{\varepsilon}_2(p)]$$

so that

$$C_\varepsilon(p) = \frac{Q(p)}{P(p)}C_\sigma(p) = 2G(p)C_\sigma(p) \tag{2.100}$$

where $G(p)$ is the operational shear modulus.

Thus, the birefringence-stress and-strain coefficients are not independent, but for the case of <u>aligned axes</u> they are uniquely related in the above manner through the time-temperature dependent shear modulus of the linear viscoelastic material.

Application of the Laplace transforms to Eqs (2.91), (2.93), and (2.98) for conditions of unit step mechanical inputs and introduction of the relation (2.100) lead to the following relationships between operational parameters and the transforms of the material properties for the specific processes of stress relaxation and strain creep :

$$
\begin{aligned}
&(1) \qquad G(p) = p\bar{G}_{rel}(p)\\
&(2) \qquad I(p) = p\bar{I}_{crp}(p)\\
&(3) \qquad C_{\varepsilon}(p) = p\bar{C}_{rel}(p)\\
&(4) \qquad C_{\sigma}(p) = p\bar{C}_{crp}(p) \, .
\end{aligned}
\qquad (2.101)
$$

We are able now to solve for $C_{\sigma_{crp}}$. Combining Eqs. (2.101.1),(2.101.3) and (2.101.4)with (2.100) we get :

$$\bar{C}_{\varepsilon_{rel}}(p) = 2p\bar{G}_{rel}(p)\bar{C}_{\sigma_{crp}}(p) \qquad (2.102)$$

Substituting Eqs.(2.101.1) and (2.101.2) in (2.102) gives :

$$\bar{C}_{\sigma_{crp}}(p) = \frac{1}{2}\bar{C}_{\varepsilon_{rel}}(p)p\bar{I}_{crp}(p) \, . \qquad (2.103)$$

After applying the convolution inversion integral to Eq. (2.103) we obtain the solution in the form :

$$(2.104)\quad C_{\sigma_{crp}}(t) = \frac{1}{2}\left\{I_{crp}(0)C_{\varepsilon_{rel}}(t) + \int_0^t C_{\varepsilon_{rel}}(t-\tau)\frac{dI_{crp}(\tau)}{d\tau}d\tau\right\} .$$

The behaviour of $I_{crp}(t)$ is required, and it can be obtained from $E_{rel}(t)$ by using the collocation method :

$$(2.105)\qquad E_{rel}(t) = E_e + \sum_{i=1}^{n} E_i e^{-\frac{t}{\tau_i}} .$$

This series can be Laplace transformed and the value of $E(p)$ and hence

$$\bar{D}_{crp}(p) = \frac{D(p)}{p} = \frac{1}{pE(p)}$$

determined. Assuming a similar type of series representation for $D_{crp}(t)$, $\bar{D}_{crp}(p)$ can be inverted by another collocation to give $D_{crp}(t)$. If compressibility is assumed, $I_{crp}(t) = 3D_{crp}(t)$.

The simplest conversion for $I_{crp}(t)$ is to use the approximation

$$(2.106)\qquad D_{crp}(t) \approx \frac{1}{E_{rel}(t)} .$$

2.5.2.3. Conversion to Inverse Relaxation Coefficient.

The equation defining the time-dependent model stress-birefringence operator C_σ^{-1}:

$$\sigma_1 - \sigma_2 = C_\sigma^{-1} n \tag{2.107}$$

Laplace transformation gives :

$$\bar{\sigma}_1(p) - \bar{\sigma}_2(p) = C_\sigma^{-1}(p)\bar{n}(p) \tag{2.107'}$$

where $C_\sigma^{-1}(p)$ may be termed the operational stress-birefringence coefficient in analogy to operational modulus.

By using a relaxation calibration test, a relation similar to Eq. (2.101.4) can be derived :

$$C_\sigma^{-1}(p) = \dot{p}\, C_{\sigma_{rel}}^{-1}(p) \tag{2.108}$$

defining a relaxation stress-birefringence coefficent. The combination of (2.108) and (2.101.4) gives :

$$\bar{C}_{\sigma_{rel}}^{-1}(p) = \frac{1}{p^2\, \bar{C}_{\sigma_{crp}}(p)} . \tag{2.109}$$

By collocation method it is possible now to obtain

$$C_{\sigma_{rel}}^{-1}(t) .$$

2.5.2.4. Conversion to Inverse Creep Coefficient.

A similar derivation for the strain variation for a step-fringe input gives a C_ε^{-1} relation similar to Eq. (2.101.3) :

$$\bar{C}^{-1}_{\varepsilon_{crp}}(p) = \frac{1}{p^2 \bar{C}_{\sigma_{rel}}(p)} \tag{2.110}$$

which can also be inverted by collocation method.

As derivated in Williams-Arenz's paper[23], the suitable material used in experiments, was Hysol 8705. This material was tested by a specimen in a special Instrom Model TT-C constant-strain-rate test machine. The specimen was in a polariscope (inside a temperature control chambre) and the fringe-order variation was obtained by recording the intensity of light through the analyzer using a photomultiplier tube equipped with suitable amplifying and graphic read-out equipment.

The data were then shifted to obtain the a_T curve from which the normalized experimental results were prepared. The relaxation modulus and the strain-optic relaxation characteristics are determined now by using Eqs. (2.85) and (2.97).

2.5.2.5. Fringe, Stress, and Strain Calculation.

After having the mechanical and optical characterization of the material, the photoviscoelastic analysis has to performed. There are two possibilities now :

(1) Analytically or experimentally knwon stresses or strains impose on a specimen and calculate the resulting fringe variation(this approach is especially suitable in comparing theoretically predicted stress fields with measured ones in cases where the experimentalist prefers not to obtain all of the isochromatics, isoclinics ans isopachics in order to calculate a single stress). Thus, the theoretician would carry through his work to the calculation of fringes, using the appropriate material-property data, for direct comparison with the experimental fringe records.

When the stresses are known and the fringes are desired, the computation is carried out using the superposition integral applied to either the stress or strain difference ; for stress it is :

$$n(t) = \frac{h_{model}}{h_{test}}\Big\{\big[\sigma_1(0) - \sigma_2(0)\big]C_{\sigma_{crp}}(t) + \\ + \int_0^t C_{\sigma_{crp}}(t-\tau)\frac{d}{d\tau}\big[\sigma_1(\tau) - \sigma_2(\tau)\big]d\tau\Big\} . \tag{2.111}$$

(2) The reverse situation may be required, particularly for situations wherein analytic solutions are unavailable. In this case the invers operators are used in conjunction with the measured fringe variation to predict the stress or strain at a given point. We use then the relation :

$$(2.112)\quad \sigma_1(t) - \sigma_2(t) = \frac{h_{test}}{h_{model}} \left\{ n(0)\, C^{-1}_{\sigma_{rel}}(t) + \int_0^t C^{-1}_{\sigma_{rel}}(t-\tau) \frac{dn(\tau)}{d\tau} d\tau \right\}$$

or, for principal strain difference :

$$(2.113)\quad \varepsilon_1(t) - \varepsilon_2(t) = \frac{h_{test}}{h_{model}} \left\{ n(0)\, C^{-1}_{\varepsilon_{crp}}(t) + \int_0^t C^{-1}_{\varepsilon_{crp}}(t-\tau) \frac{dn(\tau)}{d\tau} d\tau \right\}.$$

E.H. Dill[25] gave, as it was described in Ch. 2.4.4, the analysis of Photoviscoelasticity the general possibility of the experimental verification of results.

A very exhaustive engineering analysis of Photoviscoelasticity as presented by Williams and Arenz[23] has the accentuation to the experimental aspect of the problem, especially in connection with the application to dynamical problems.

Many papers being published recently (References 24-33) are concerned to the technology of materials being applied to different fields of photoviscoelasticity in both static and dynamic cases, very often in connection with important engineering problems.

2. 6. Some Recent Investigations in Photoviscoelasticity.

Isaac M. Daniel [26] has investigated the quasi-static viscoelastic properties of the plasticized polyvinil chloride because of its low modulus, transparency, birefringence and viscoelasticity. He tested it by both, creep and relaxation testing using the Moiré method of strain analysis for continuous recording of axial and transversal deformations in a creep specimen. The simultaneous recording of these two deformations affords the separation of deviatoric and dilatational effects, provided superposition of these two effects is valid.

In the creep test, where stress was maintained constant, the birefringence can be decomposed into a constant part due to stress and a time-varying part due to strain, as follows :

$$n(t) = n_\sigma + n_\varepsilon(t) .$$

In the relaxation test, the birefringence was analyzed also in two parts, one constant with time due to strain and one varying with time due to stress,

$$n(t) = n_\varepsilon + n_\sigma(t)$$

and by both tests the stress-fringe value $\bar{f}_\sigma$ and the strain fringe value $\bar{f}_\varepsilon$ were obtained.

I. M. Daniel[27] investigated also dynamical properties of the viscoelastic plasticized polyvinil chloride.

Dynamical, mechanical and optical properties can be determined indirectly from quasi-static experiments at different temperatures using the temperature-time equivalence principle (Theocaris, Mylonas, Williams). At the Daniel's study these properties were examined directly, dynamically on the basis of the following theoretical considerations :

Mechanical properties : It is necessary to consider the dynamical behaviour of visçoelastic material by studying the response to sinusoidally varying stress or strain. If a sinusoidally stress

$$s_{ij} = s_0 \sin \omega t \tag{2.114}$$

is applied to material, the strain will vary with the time according to the relation

$$e_{ij} = e_0 \sin(\omega t - \delta) \tag{2.115}$$

i.e. with a phase difference δ where s_{ij} and e_{ij} are stress, resp. strain deviators. The stress-strain ratio is expressed by the complex shear modulus :

$$G^*(i\omega) = \frac{s_{ij}}{2e_{ij}} = G_1(\omega) + iG_2(\omega) \qquad (2.116)$$

with

$$G_1(\omega) = \frac{s_0}{2e_0}\cos\delta; \qquad G_2(\omega) = \frac{s_0}{2e_0}\sin\delta. \qquad (2.117)$$

The real part is called the storage modulus, and the imaginary part G_2 the loss modulus.

The tested specimen was a cantilever beam subjected to sinusoidal motion at the free end, and the data recorded were acceleration(related to displacement) at the free end, and force (related to stress) at the fixed end (Fig. 29). Of course, both stress and displacement at both ends are required to define uniquely a complex modulus. At the experiment, because of the existing boundary conditions, the approximate value of the complex modulus was obtained. This value can be corrected by successive approximations.

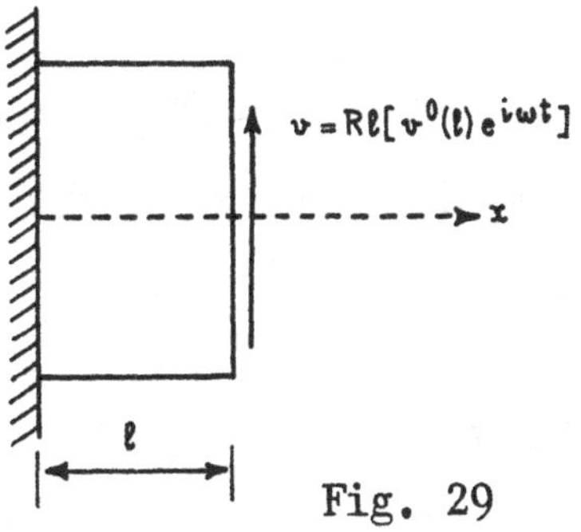

Fig. 29

Let :

$$(2.118)\qquad \begin{aligned} s(x,t) &= R\ell\left[s^0(x)e^{i\omega t}\right] \qquad &(a)\\ e(x,t) &= R\ell\left[e^0(x)e^{i\omega t}\right] \qquad &(b)\\ v(x,t) &= R\ell\left[v^0(x)e^{i\omega t}\right] \qquad &(c) \end{aligned}$$

be shear stress, strain and displacement for a simple-shear specimen, where $s^0(x)$, $e^0(x)$, and $v^0(x)$ are in general complex numbers.

The equation of motion :

$$(2.119)\qquad \frac{\partial s^0(x)}{\partial x} = -\rho\omega^2 v^0(x) .$$

The strain-displacement relation :

$$(2.120)\qquad e^0(x) = \frac{1}{2}\frac{\partial v^0(x)}{\partial x} .$$

The stress-strain relation :

$$(2.121)\qquad s^0(x) = 2G^*(i\omega)e^0(x) .$$

The data obtained experimentally are the stress at the fixed end and the displacement at the free end.

$$(2.122)\qquad \begin{aligned} s(0,t) &= A_0\cos(\omega t + \delta) \qquad &(a)\\ v(\ell,t) &= B_\ell \cos\omega t . \qquad &(b) \end{aligned}$$

Substituting Eqs. (2.120) and (2.121) in (2.119) we obtain

$$\frac{\partial^2 v^0(x)}{\partial x^2} + q^2 v^0(x) = 0 \tag{2.123}$$

where

$$q^2 = \frac{\varrho\omega^2}{G^*(i\omega)} .$$

The solution of (2.123) satisfying the boundary condition $v^0(0) = 0$ is

$$v^0(x) = C \sin qx .$$

Eqs. (2.122b) and (2.118c) give $C = B_\ell/(\sin q\ell)$,i.e.

$$e^0(x) = \frac{B_e q \cos qx}{2 \sin q\ell} . \tag{2.124}$$

Thus, the complex shear modulus is given by

$$G^*(i\omega) = \frac{s^0(0)}{2e^0(0)} = \frac{A_0 e^{i\delta}}{B_\ell/\ell} \frac{\sin q\ell}{q\ell} \tag{2.125}$$

(q is a function of the complex modulus, thus, Eq. (2.125) is an implicit function). The first approximation obtained directly from the experimental data is

$$\frac{A_0 e^{i\delta}}{B_\ell/\ell} .$$

This value of G^* can be used for the computation of q

in Eq. (2.125), and thus a second approximation for G^* is obtained.

Optical Properties : Optical properties were studied by Coker and Filon[1], Mindlin[20], Read[21], Daniel[27] and others. Dill[25] proposed the stress-optic law for viscoelastic materials in the integral form in which the birefringent properties of the material are charaterized by a time-dependent stress-fringe value $f_\sigma(t)$. The difference of normal stresses :

$$(2.126) \quad \sigma_{xx} - \sigma_{yy} = \frac{2}{h}\int_0^t f_\sigma(t-\tau)\frac{d}{d\tau}\left[(n_1 - n_2)\cos 2\Phi_n\right]d\tau$$

and the shear stress :

$$(2.126) \quad \sigma_{xy} = \frac{1}{h}\int_0^t f_\sigma(t-\tau)\frac{d}{d\tau}\left[(n_1 - n_2)\sin 2\Phi_n\right]d\tau .$$

The fringe value f_σ is obtainable by a photocreep test (for quasi-static applications) or from sinusoidal oscillation test (for dynamic applications).

In a sinusoidal oscillation test, stress and birefringence may in general be out of phase. The complex stress-fringe value as a function of frequency can be defined as

$$f_\sigma^*(i\omega) = f_{\sigma_1}(\omega) + i f_{\sigma_2}(\omega) . \qquad (2.127)$$

This function of frequency can be converted into a function of time by means of exact or approximate interrelations for use in the stress-optic expressions (2.126). Another form of the stress-optic law, proposed by I. Daniel, using the complex fringe value directly is :

$$\sigma_{xx} - \sigma_{yy} = \frac{2}{h}\int_0^\infty f_\sigma^*(i\omega)(\bar{n}_{xx} - \bar{n}_{yy})e^{i\omega}\, d\omega \qquad (2.128)$$

and

$$\sigma_{xy} = \frac{1}{h}\int_0^\infty f_\sigma^*(i\omega)\bar{n}_{xy}e^{i\omega}\, d\omega \qquad (2.129)$$

where

$$\bar{n}_{ij} = \frac{1}{\pi}\int_0^\infty n_{ij}e^{i\omega t}dt .$$

The dynamic properties of viscoelastic materials in shear were studied by R. L. Adkins [36] obtaining the value of complex shear modulus. He proposed the complete equipment to this purpose.

A. S. Miguel and R.H. Silver [37] constructed a very suitable normal-incidence reflective polariscope capable of measuring isoclinics and isochromatics as a function of time. The instrument consists of eight unit polariscopes, relatively oriented, which are indexed into position within a short period of time to obtain a set of eight data.

J. T. Pindera [28] discussed the problem of physical similarity between model and object at viscoelastic materials in the case when the model material behaves in a certain range as a linear viscoelastic material. The reliability of the common forms of specimens (tensile, bending, etc) is discussed from the point of view of the existency of creep and relaxation phenomena and it is shown how to obtain sufficiently accurate relations between stresses, strain, birefringence and time using tapered specimens.

R.M. Hackett and E.M. Krokosky [32] proposed a technique of photoviscoelastic stress analysis applied to an idealized model of polyphase-material system having a viscoelastic binder in order to determine the time-dependent stress redistribution.

V. Brčić and M. Nešović [30] used a low-modulus material on the gelatine basis, exhibiting viscoelastic behaviour and possessing extremely high optical sensitivity and extremely low modulus of elasticity, to simulate body forces due to the own weight at some high dam models.

H.F. Brinson [33] studied the mechanical and optical characterization of Hysol 4290 as a function of time and temperature. He discussed the kinetic theory of polymers, the WLF-equation and the time-temperature superposition principle.

The properties of Hysol 4290 are strongly time and temperature dependent, especially in the transition range of the material. Studying this problem, Theocaris [21] (epoxy resins) and Williams [34] (polyurethane rubber Hysol 8705) used the time-temperature superposition principle.

The high polymers, whose properties vary widely with temperature, have three characteristic regions : the glassy region, the transition region and the rubbery region. The most interesting is the transition region.

Viscoelastic properties of polymers are tested by creep and relaxation test, usually by uniaxial tensile specimens. If a constant force is instantaneously placed on a specimen, the resulting strain and birefringence are :

$$\begin{aligned} \varepsilon(t) &= \sigma_0 D(t) ; \quad \sigma(t) = \sigma_0 H(t) \quad t > 0 \\ n(t) &= \sigma_0 C_\sigma(t) ; \quad \sigma(t) = \sigma_0 H(t) \quad t > 0 \end{aligned} \tag{2.130}$$

where

$D(t)$ is the creep compliance,

$C_\sigma(t)$ is the stress-optic coefficient,

$H(t)$ is the Heaviside step function.

For relaxation the expressions are similar, changing the strain with stress and reversely.

For the linear case, the Boltzmann's superposition principle is applicable and the last two equations can be generalized to :

$$\begin{aligned} \varepsilon(t) &= \int_0^t D(t-\tau) \frac{d\sigma(\tau)}{d\tau} d\tau \\ n(t) &= \int_0^t C_\sigma(t-\tau) \frac{d\sigma(\tau)}{d\tau} d\tau . \end{aligned} \tag{2.131}$$

These equations hold if the axes of principal strain and stresses are coincident with the principal axes of polarization. This can exist for uniaxial tension.

There exist also the relation between the creep compliance, $D(t)$, and the relaxation modulus, $E(t)$:

$$\int_0^t E(t-\tau) D(\tau) d\tau = t \tag{2.132}$$

or, approximately :

$$E(t) \approx \frac{1}{D(t)} . \qquad (2.132a)$$

The characterization of the mechanical properties of polymers can be also defined by the use of mechanical models. For a generalized Maxwell's model, e.g., the relaxation modulus can be written as

$$E(t) = \sum_{i=1}^{n} (E_i e^{-\frac{t}{\tau_i}})$$

where n is the number of parallel Maxwell elements, E_i is the modulus of each spring, and τ_i is the relaxation time for each Maxwell model. However, it is generally more convenient to use a continuous spectrum of relaxation times $(n \rightarrow \infty)$ and to write the last equation in integral form, taking $\ln \tau$ as the variable of integration :

$$E(t) = \int_{-\infty}^{+\infty} H(\ln\tau) e^{-\frac{t}{\tau}} d(\ln\tau) \qquad (2.133)$$

where $H(\ln\tau)$ is defined as a distribution function of relaxation times, i.e. a relaxation spectrum. The expression $e^{-\frac{t}{\tau}}$ is referred to as an intensity function.

2.7. The Time-Temperature Superposition Principle. [23]

The time-temperature superposition principle was originally developed on empirical basis, but it can be logically developed from the kinetic theory of polymers. In the kinetic theory, time and temperature have a very important role. Eq. (2.133) is a function of both temperature and time and should be properly written in the form

$$E(t,T) = \int_{-\infty}^{+\infty} H(\ln\tau,T)\,e^{-\frac{t}{\tau}}\,d(\ln\tau)\,. \tag{2.133a}$$

The temperature dependence of the relaxation spectrum can be separated from the time dependence resulting in Eq. (2.133a) being rewritten as

$$E(t,T) = \varrho T\int_{-\infty}^{+\infty} h(\ln\tau)\,e^{-\frac{t}{\tau}}\,d(\ln\tau) \tag{2.134}$$

where ϱ is the density, T the temperature. In addition, the kinetic theory shows that the ratio of the relaxation times at two different temperatures can be expressed as,

$$a_T = \frac{\tau}{\tau_0} = \left(\frac{\xi}{\xi_0}\right)\frac{\varrho_0 T_0}{\varrho T} \tag{2.135}$$

where a_T is defined as a shift function, and ξ is defined as the viscosity of the polymer. From Eqs.

(2.134) and (2.135) it follows

$$E(t',T_0) = \frac{\varrho_0 T_0}{\varrho T} E(t = a_T t',T). \qquad (2.136)$$

This is a statement of the time-temperature superposition principle and indicates that the relaxation modulus at temperature T_0 and time t' can be obtained from the relaxation modulus measured at temperature T and time t. On a plot of modulus vs. logarithmic time, this amount to a vertical shift of the magnitude $\varrho_0 T_0/(\varrho T)$, and a horizontal shift of the magnitude $\log a_T$. Equation (2.136) is applicable to optical as well as mechanical data. Williams and Arenz[23] compare the experimentally determined shift function to the WLF equation

$$\log a_T = -\frac{K_1 \frac{T}{T_0}}{K_2 + (T - T_0)} \qquad (2.137)$$

where K_1 and K_2 are constants depending on the reference temperature T_0 . If the glas transition temperature, T_g , is used as the reference temperature, the constants have the values : $K_1 = 17.44$; $K_2 = 51.6$. Equation (2.137) is only valid from the T_g upon to approximatelly $T_g + 100\,°C$. The kinetic theory should be applicable if the measured shift function satisfies Eq. (2.137).

H.F. Brinson investigated the properties of Hysol 4290 by a series of creep and relaxation tests, comparing different results related to the dependence of compliance with temperature and time, and comparing his experimental data to the WLF-equation.

P. Theocaris[38], discussed also the time-temperature superposition principle by analyzing the photoviscoelatic behaviour of epoxy polymers.

He noticed it was established experimentally, that the retardation or relaxation times constituting the viscoelastic spectra of the polymers decrease rapidly with increasing temperature. In order to avoid the complicate temperature-dependent relationship is simpler to introduce this dependence in the viscoelastic behaviour by a method of reduced variable time or frequency based on the time-temperature superposition principle. According to this principle, viscoelastic data at one temperature are transformed to another temperature by a simple multiplicative transformation of the time scale. If the characteristic functions are plotted in a log time scale, this transformation degenerates to a mere parallel shift of the log time scale (see Fig. 28). Due to this, the creep curves plotted vs. log time are identical at difference temperatures save for a shift in the origin of the log time axis. The mathematical form of this principle, proposed by Ferry :

$$\frac{\varrho_0 T_0}{\varrho T} E_T\left\{a_T(\tau)\right\} = E_{T_0}(\tau)$$

(this is the before deduced Eq. (2.136)). Here is $E_T(\tau)$ the distribution of extension relaxation times at a temperature T, $E_{T_0}(\tau)$ the same at T_0 ; a_T is a function of temperature only, ϱ_0 and ϱ densities at T_0 and T, and T_0/T is a correction factor.

The time-temperature superposition principle was shown to hold for the optical properties of high polymers.

P. Theocaris investigated the behaviour of a series of different kinds of epoxy polymers with different percentage of plasticizer and he established that addition of plasticizer results in lowering the rubbery-state temperature of each polymer.

P. Theocaris and C. Mylonas [24] investigated the viscoelastic behaviour of birefringent coatings by creep and relaxation tests using different kind or pure epoxy resins. Their results can be summarized as follows :

Pure epoxy resins presented insignificant creep and relaxation, but it can be intensified with increasing amount of plasticizer.

The stress-strain-optical relations of all resins are linear and are identical in tension, compression and pure shear.

In the creep tests, the birefringence and the strain at the time t after loading could be expressed in the form

$$f_t = f_\sigma + at + b\log t \tag{2.138}$$

where f_σ , a and b are parameters depending on the material properties and experimental conditions.

The relaxation law can be expressed as a logarithmic function of the time t.

All these material are suitable for use as birefringent coatings when the interface strains are given by a "product function":

$$u_i = \bar{u}_i g(t) .$$

If not, the photoelastic measurements must be made after a relatively long time following each strain variation.

2. 8. Concluding Remarks. [38]

As mentioned earlier, the linearity of viscoelastic materials (high polymers) is pronounced by the fact that the time-dependent functions are approximately linearly proportional to the applied constant stresses or strains. Many photoelastic materials are linearly viscoelastic.

Departures from linearity are observed for materials stretched beyond a certain limit of deformation in their rubbery state. Moreover, some polymers in their glassy stage present nonlinear phenomena even when the strains are relatively small. The treatment of this linearity is different from that in the rubbery domain, where any deviation from Hooke's law is associated with large strains. The theory of nonlinear photoviscoelasticity has been still in its developmental stage.

The linear photoviscoelastic behaviour of viscoelastic materials simplifies considerably an exact photoelastic analysis and allow their use along the whole range of time and temperature.

The application of Laplace transforms has been giving a successful perspective to the solution of many important problems of viscoelasticity by reducing them to the associated problems of elasticity. Inversion of the solution of the associated elastic problem to the original plane provides the viscoelastic analysis.

Many approximate methods of the inversion of Laplace transforms have been proposed lately by Alfrey[39], Shapery[35], and others. The analytic difficulties encountered in the solutions of problems containing transient nonuniform temperature fields and moving boundaries of linear viscoelastic materials become more pronounced. This is because the introduction of the temperature dependence excludes the application of Laplace transforms necessary to eliminate the time dependence of the problem, since the basic equations relating the time and temperature effects no longer maintain their convolution form.

There is a series of methods where viscoelastic problems are attacked by approximate numerical methods without introducing Laplace or Fourier transforms.

<u>A Review on Basic Linear Relations at Viscoelastic Materials[38]</u>. The behaviour of photoviscoelastic materials is defined by three types of compliances and moduli corresponding to some characteristic modes of loading :

The extension, the shear and the bulk compliances : $D(t)$, $I(t)$, $B(t)$;
moduli : $E(t)$, $G(t)$, $K(t)$.

The nature of their time-dependence is one of the basic problems of the phenomenological analysis of viscoelasticity.

These quantities describe the <u>mechanical</u> behaviour of material. The <u>optical</u> properties are related to variation of birefringence with stress and strain through the stress-optical and strain-optical coefficients in creep and relaxation, $C_\sigma(t)$ and $C_\varepsilon(t)$.

The linear behaviour of viscoelastic materials may be expressed by laws maintaining linear differential operators which, when properly choosen, may reproduce any arbitrary creep or relaxation function for any time-dependent loading. The different types of rheological models are applicable here successfully.

A more general representation in term of creep compliances or relaxation moduli functions can be given by viscoelastic operators in integral form. The following relations exist : (see page 152)

$$(2.139)\quad \begin{cases} \varepsilon_i(t) = \int_0^t D(t-\tau)\frac{d\sigma_i(t)}{d\tau}d\tau \\ \sigma_i(t) = \int_0^t E(t-\tau)\frac{d\varepsilon_i(t)}{d\tau}d\tau \end{cases}$$

$$(2.140)\quad \begin{cases} e_{ij}(t) = \frac{1}{2}\int_0^t I(t-\tau)\frac{ds_{ij}(\tau)}{d}d\tau \\ s_{ij}(t) = 2\int_0^t G(t-\tau)\frac{de_{ij}(\tau)}{d}d\tau \end{cases}$$

$$(2.141)\quad \begin{cases} e_{ii}(t) = \frac{1}{3}\int_0^t B(t-\tau)\frac{d\sigma_{ii}(\tau)}{d\tau}d\tau \\ \sigma_{ii}(t) = 3\int_0^t K(t-\tau)\frac{d\varepsilon_{ii}(\tau)}{d\tau}d\tau \end{cases}$$

$$(2.142)\quad \begin{cases} \gamma_m(t) = \int_0^t C_\varepsilon(t-\tau)\frac{d}{d\tau}\left[N_\varepsilon(\tau)\cos 2\vartheta(\tau)\right]d\tau \\ 2\sigma_m(t) = \int_0^t C_\sigma(t-\tau)\frac{d}{d\tau}\left[N_\sigma(\tau)\cos 2\vartheta(\tau)\right]d\tau \end{cases}$$

where σ_m , γ_m are maximum shear stress and strain resp., N_σ , N_ε birefringence due to an applied maximum shear stress or strain, and $\vartheta(\tau)$ the angle between the principal axes of birefringence and the principal axes of stress and strain.

The relation between the compliances and the corresponding moduli :

$$\int_0^t E(t-\tau)D(\tau)d\tau = t$$
$$\int_0^t G(t-\tau)I(\tau)d\tau = t \qquad (2.143)$$
$$\int_0^t K(t-\tau)B(\tau)d\tau = t\,.$$

The stress-and strain-optical coefficients C_σ and C_ε are uniquely interrelated in the case of alignment of mechanical and optical principal axes through the time-dependent shear compliance and relaxation modulus.

Eqs. (2.139) - (2.142) can be Laplace transformed, f.i. :

$$\sigma_i(p) = \bar{E}(p)\bar{\varepsilon}_i(p)$$
$$2\bar{\sigma}_m(p) = \bar{C}_\sigma(p)\bar{N}_\sigma(p)\cos 2\vartheta(p) \qquad (2.144)$$

etc.

These relations show the quasi-static character of the stress-strain-optical laws in transform plane and they enable to interpret any viscoelastic solution as a group of elastic solutions by considering the elastic constants and the boundary conditions to be appropriate functions of the parameter p.

The lateral contraction $\nu(t)$ of the material is defined by the relations :

$$-\varepsilon_y(t) = \int_0^t \nu_c(t-\tau)D(t-\tau)\frac{d\sigma_z(\tau)}{d\tau}d\tau \tag{2.145}$$
$$-\varepsilon_y(t) = \int_0^t \nu_r(t-\tau)\frac{d\varepsilon_z(\tau)}{d\tau}d\tau$$

where z is the longitudinal axis. The function $\nu_c(t)$ defined from a creep test $[\sigma(t) = \sigma_0]$ is in principle different than the function $\nu_r(t)$ defined from a relaxation test. The creep and relaxation lateral-contraction ratios may be expressed by the relations :

$$\nu_c(t) = \left[\frac{1}{2}\frac{I_0}{D(t)} - 1\right] + \frac{1}{2}\int_0^t \frac{dI(t-\tau)}{d\tau}\frac{1}{D(\tau)}d\tau \tag{2.146}$$
$$\nu_r(t) = \left[\frac{1}{2}\frac{E_0}{G(t)} - 1\right] + \frac{1}{2}\int_0^t \frac{dE(t-\tau)}{d\tau}\frac{1}{G(\tau)}d\tau .$$

Between these two lateral-contraction ratios holds the relation :

$$\int_0^t \nu_r(t-\tau)\nu_c^{-1}(\tau)\,d\tau = t . \tag{2.147}$$

Eqs. (2.146) can be also written in the transform

plane :

$$\bar{\nu}_c(p) = \left[\frac{1}{2}\frac{\bar{I}(p)}{\bar{D}(p)} - 1\right]; \qquad \nu_r(p) = \left[\frac{1}{2}\frac{\bar{E}(p)}{\bar{G}(p)} - 1\right]. \qquad (2.148)$$

These equations may be used for the determination of the lateral-contraction functions in terms of the measured compliance and modulus functions in the transform plane. This process is the analogy to the determination of relations between elastic constants.

The alignment of mechanical and optical axes exists in the glassy and rubbery stage, and only during transition region of the material the axes are not aligned.

The incompressibility assumption is valid only in the rubbery region of loading. This assumption, together with the alignment of axes, reduces the independent material functions needed to characterize the viscoelastic behaviour of the material to one function, i.e. any one of the relaxation moduli or creep compliances (extension, shear, or bulk). However, it is accepted that a pair of independent characteristic functions must be known for the complete determination of the mechanical properties of a linear viscoelastic material, in complete analogy with the two independent elastic constants of isotropic elasticity. Another function is needed for the characterization of optical properties of the material which will relate birefringence with stress or strain.

P. Theocaris [38] showed that the static pure-tension test in creep or relaxation is sufficient for the complete determination of the viscoelastic behaviour of the materials, provided that only a generic value of another characteristic function is known along the response spectrum of the material. On this basis he proposed a step-by-step method to determine the whole viscoelastic spectrum of the considered material [38].

Chapter 3.
Photothermoelasticity.

3. 1. Introduction

Photothermoelasticity is an interdisciplinary field[44] utilizing photoelasticity, thermoelasticity and photothermography, determining the thermal-stress fields by means of photoelasticity. The external load is here represented by temperature gradients.

E.E. Weibel[40] (1938) was the first who applied Photoelasticity to the solution of thermal stress problems. However, Gerard, Gilbert and Tramposch[41]-[44] established it as a reliable method in this field. They used an imbedded polariscope to isolate a single internal plane of the model for study, thus applying what are essentially two-dimensional techniques to afford full-field studies of internal stresses on selected planes in three-dimensional thermally stressed models.

The problems which have been studied and solved can be divided into three categories : shrinkage stresses resulting from a uniform change in temperature in structures composed of materials with different coefficient of thermal expansion ; stresses in multiply connected cylinders subjected to steady-state linear temperature gradients ; and transient thermal stresses including thermal shock.

The main photoelastic methods used were : standard transmission polariscope on two-dimensional models, reflection techniques to study surface stresses on large models, a combination of mechanical prestraining and three-dimensional stress freezing techniques, and built-in polariscope in three-dimensional models.

Since photoelastic materials tend to creep at elevated temperatures, temperature gradient is generally established by working below room temperature. Since the obtained fringe patterns representing the thermal stress fields are closely with the conditions at the specimen boundaries, the important task concerned with the problem of photothermoelasticity is the simulation and measurement of the corresponding thermal fields.

The thermal expansion components of the thermoelastic strains produce an isotropic field, thus, they do not produce any effects upon the fringe patterns. Therefore, it is not necessary to introduce corrections to the observed results as may be required with other experimental thermal stress methods.

Because the experimental thermal conditions are usually below ambient, it is necessary to investigate the variations in some physical quantities, modulus of elasticity E, coefficient of thermal expansion α , and material fringe value f, over a broad temperature range, from - 40°C to + 80°C. It is necessary to have the materials which posses high optical sensitivity. A convenient way to evaluate this sensitivity is the so-called "<u>figure of merit</u>" defined by the expression

$$Q_\alpha = \frac{\alpha E}{f} .$$

The higher the value of Q_α the more sensitive is the material fringe wise for a given value of αE .

The first works in photothermoelasticity were connected with some simple problems, like beam in bending suddenly cooled by dry ice on one edge, then a disc with an embedded steel plug in the center, all in a uniform thermal field.

In structural parts,thermal stresses can be obtained either by thermal-gradient field, or by a uniform temperature field but by interference between two different materials.

The first theory of thermoelasticity was given by M.C. Duhamel (1837) who proposed the well-known equation of the linear theory of thermoelasticity :

$$(3.1) \qquad kT_{,ii} = \varrho c\dot{T}\left[1 - \frac{3\lambda + 2\mu}{\varrho c}\alpha^2 T_0\left(\frac{\dot{\varepsilon}_{kk}}{\alpha\dot{T}}\right)\right]$$

where is

k thermal conductivity

ϱ density

c specific heat

T temperature

α thermal expansion coefficient

λ, μ Lamé's constants.

This is the equation of the coupled theory of thermoelasticity.

The first problem of heat conduction, in the case without heat sources or sinks, is defined by the equation

$$(3.2) \qquad kT_{,ii} = \varrho c\dot{T} .$$

After the temperature field in the body being determined by solving this equation, i.e. the temperature T as known function of coordinates, the theory of the linear uncoupled thermoelasticity has to be applied.

3. 2. Similarity Relations. [45]

In the following text the derivation of similarity relations for time, temperature, displacements, strain and stresses between model and prototype, for the case of uncoupled, quasi-static thermoelastic problem, will be given. It is assumed that all mechanical, optical and thermal constants remain invariant under small temperature change [15].

By using the well-known relations: the equilibrium equations (body forces neglected)

$$\sigma_{ij,j} + f_i = 0 \tag{3.3}$$

the stress-strain relations (the Hooke's material is supposed):

$$\sigma_{ij} = \delta_{ij}\lambda\varepsilon_{kk} + 2\mu\varepsilon_{ij} - \delta_{ij}(3\lambda + 2\mu)\alpha T \tag{3.4}$$

the strain-displacement relations :

$$\varepsilon_{ij} = \frac{1}{2}(u_{i,j} + u_{j,i}) \tag{3.5}$$

(small deformations are assumed), we obtain the well-known Navier's equations :

$$u_{i,jj} + \frac{1}{1 - 2\nu}u_{j,ji} - \frac{2(1 + \nu)}{1 - 2\nu}\alpha T_{,i} = 0. \tag{3.6}$$

Eq. (3.6) related to prototype. The same equation will be written for model too, by using subscript or superscript "m" :

(3.7) $$u^m_{i,j'j'} + \frac{1}{1-2\nu} u^m_{j,j'i'} - \frac{2(1+\nu_m)}{1-2\nu_m} \alpha_m T_{,i} = 0$$

(where primed indices refer to model coordinates).

The displacements u_i and u^m_i can be expressed in terms of the thermo-elastic-displacement potential function Φ and Φ^m, such that

(3.8) $$u_i = \Phi_{,i} \; ; \qquad u^m_i = \Phi^m_{,i'} \, .$$

Now, we introduce the following scaling terms :

(3.9)
$$\delta = T/T^m \text{ (temperature scaling)}$$
$$\lambda = X_i/X^m_i \text{ (geometric scaling)}$$
$$\gamma = u_i/u^m_i \text{ (displacement scaling).}$$

Eqs. (3.8) and (3.9) give after integration :

(3.10) $$\Phi = \lambda\gamma\Phi^m .$$

Also, from the last two terms of Eqs. (3.9) it can be given

(3.11) $$u_{i,j} = \lambda\gamma u^m_{i,j} \, .$$

Introducing Eqs. (3.8), (3.9) and (3.10) into Eqs.

(3.6) and (3.7) and integrating, we have

$$\phi_{,jj} - \frac{1+\nu}{1-\nu}\alpha T = 0 \tag{3.12a}$$

or

$$u_{j,j} = \frac{1+\nu}{1-\nu}\alpha T \tag{3.12b}$$

and

$$\phi_{,jj} - \frac{1+\nu_m}{1-\nu_m}\frac{\alpha_m \gamma}{\delta\lambda} T = 0 \tag{3.13a}$$

or

$$u_{j,j} = \frac{1+\nu_m}{1-\nu_m}\frac{\alpha_m \gamma}{\delta\lambda} T \, . \tag{3.13b}$$

By comparing (3.12b) and (3.13b) :

$$\gamma = \frac{1+\nu}{1-\nu}\frac{1+\nu_m}{1-\nu_m}\frac{\alpha}{\alpha_m}\delta\lambda \, . \tag{3.14}$$

Because of Eq. (3.5) and the relation

$$\sigma_{ij} = \frac{E}{1+\nu}\left[\frac{1}{2}(u_{i,j} + u_{j,i}) - \delta_{ij} u_{j,j}\right] \tag{3.15}$$

and by using Eqs. (3.11) and (3.14) in (3.5) and

(3.15) we obtain finally the expressions for strain and stress :

$$\varepsilon_{ij} = \frac{1+\nu}{1-\nu}\frac{1-\nu_m}{1+\nu_m}\frac{\alpha}{\alpha_m}\delta\varepsilon_{ij}^m \tag{3.16}$$

and

$$\sigma_{ij} = \frac{E}{E_m}\frac{1-\nu_m}{1-\nu}\frac{\alpha}{\alpha_m}\delta\sigma_{ij}^m . \tag{3.17}$$

For a state of <u>plane stress</u>, Eqs. (3.12b) and (3.13b) have slightly different coefficients. In this case, Eqs. (3.14), (3.16) and (3.17) become :

$$\gamma = \frac{1-\nu}{1-\nu_m}\frac{\alpha}{\alpha_m}\lambda\delta \tag{3.18}$$

$$\varepsilon_{ij} = \frac{1-\nu}{1-\nu_m}\frac{\alpha}{\alpha_m}\delta\varepsilon_{ij}^m \tag{3.19}$$

$$\sigma_{ij} = \frac{E}{E_m}\frac{\alpha}{\alpha_m}\delta\sigma_{ij}^m . \tag{3.20}$$

<u>Temperature and time similarity</u>. The thermoelastic-similarity relations depend directly upon the temperature-scaling term. Generally, model materials possess thermal properties which are quite

different from the prototype. However, similarity law relating the pertinent thermal variables are known.

By starting with the heat-conduction equation,

$$KT_{,ii} = T_{,t} \tag{3.21}$$

and defining a time scaling term as

$$\tau = \frac{t}{t_m}, \tag{3.22}$$

Fourier derived the relationship :

$$\tau = \left(\frac{K_m}{K}\right)\lambda^2 . \tag{3.23}$$

In the following text, the influence of heat sources or sinks will be shown. Fourier's heat equation including heat sources and sinks has the form

$$KT_{,ii} + Q = T_{,t} \tag{3.24}$$

and for model

$$K^m T^m_{,i'i'} + Q^m = T^m_{,t_m} \tag{3.25}$$

where Q is the heat source or sink term. Substituting Eq. (3.11), the first two terms of Eq. (3.9) and an additional scaling term defined as

$$\beta = \frac{Q}{Q_m} \tag{3.26}$$

heat source or sink scaling, into (3.24) gives :

$$\left(\frac{\delta}{\tau}\right) K^m T^m_{,i'i'} + \beta Q^m = \left(\frac{\delta}{\tau}\right) T^m_{,t_m} \tag{3.27}$$

For similarity, Eq. (3.27) must be identical to Eq. (3.25); thus

$$\beta = \frac{\delta}{\tau}$$

or, by Eq. (3.23) :

$$\beta = \frac{\delta K}{\lambda^2 K_m} . \tag{3.28}$$

The similarity relations enable the transition of the model results to the prototype. Concerning photothermoelasticity it is known that thermally induced stresses can be measured in birefringent models by the same technique used in classical photoelasticity.

According to Eq. (3.22), a time dilation occurs in the model which can slow down transient phenomena (if $\tau < 1$) making it easier to note or record the events.

Similarly by a proper choose of the coefficient $\delta = T/T_m$ it is possible to design an experiment such that the maximum temperature and stress in the model are within desired limits.

G.S. Vardanian and N.I. Prigorovski [46] proposed a method of simulating thermal stresses in arbitrary three-dimensional model by imposing frozen deformations according to the free thermal expansion of small parts from which the considered model has to be composed (according to the existing thermal field). By heating the model is defrozen, but a redistribution of deformations occurs and the model has the residual thermal stresses now. By cooling the model to the room temperature these stresses are frozen and can be investigated by conventional methods of photoelasticity.

In the general case, it is necessary to produce the frozen strains in the model parts :

$$\varepsilon_k^0 = \varepsilon_y^0 = \varepsilon_z^0 = \alpha T$$

or the stresses (3.29)

$$\sigma_x^0 = \sigma_y^0 = \sigma_z = \frac{\alpha E}{1-2\nu} T .$$

(Note : The Poisson's ratio ν has to be different from 1/2).

In two-dimensional case :

$$\varepsilon_x^0 = \varepsilon_y^0 = \alpha T ; \quad \sigma_x^0 = \sigma_y^0 = \frac{\alpha E}{1-\nu} T . \qquad (3.30)$$

The differential equation and the boundary conditions:

$$\nabla^4 \Phi + K\nabla^2 T = 0$$

(3.31)

$$\frac{\partial}{\partial s}\left(\frac{\partial \Phi}{\partial x}\right) = 0; \quad \frac{\partial}{\partial s}\left(\frac{\partial \Phi}{\partial y}\right) = 0$$

where

$K = E\alpha$ in the case of plane stress

$K = \dfrac{E\alpha}{1-\nu}$ in the case of plane strain

ane Φ is the Airy's stress function.

It can be shown that in the case of plane stress, at both simply and multiply connected domains, the thermal stresses at boundary points do not depend on ν and E. In the case of plane strain the thermal stresses in boundary points are proportional to E and $1/(1-\nu)$.

The solution of Eq. (3.31) can be obtained by using the equivalent system

(3.32) $$\nabla^2 V = T; \quad \nabla^4 U' = 0; \quad U - V = \Phi_1 = \frac{\Phi}{K}$$

and by the use of the Muskhelishvili's method of complex functions. One obtains then :

(3.33) $$\sigma_x + \sigma_y = K\left[2\left\{\varphi'(z) + \overline{\varphi'(z)}\right\} - T\right]$$

$$\sigma_x - \sigma_y + 2i\tau_{xy} = K\left[2\{\bar{z}\varphi''(z) + \Psi'(z)\} - 4\frac{\partial^2 V}{\partial z^2}\right] \quad (3.33)$$

$$2G(u + iv) = K\left[\varkappa\varphi(z) - z\overline{\varphi'(z)} - \overline{\Psi(z)} - 2\frac{\partial V}{\partial \bar{z}}\right] \quad (3.34)$$

where

u and v are displacement components

G is shear modulus

$\varkappa = (3 - \nu)/(1 + \nu)$ for plane stress

$\varkappa = 3 - 4\nu$ for plane strain

The complex functions $\varphi(z)$ and $\Psi(z)$ can be obtained from boundary conditions.

In the case of a simply connected domain the functions $\varphi(z)$ and $\Psi(z)$ are holomorphic in the considered domain the Poisson's ratio do not appear in the boundary condition, and the state of stress is not dependent on ν.

3. 3. Some Recent Papers in Photothermoelasticity.

V. Švec [47] studied the problem of photothermoelasticity by analyzing the similarity laws and by verifying the results on the example of a beam subjected to the temperature change by means of a block of dry ice. Švec studied also Biot's analogy connected with the problem of dislocations at multiply connected bodies.

Similar investigations were carried out also by Z. Orlós and Z. Dylag [48].

H. Tramposch and G.Gerard [41] - [44] published a series of papers discussing the two-and three-dimensional problems of photothermoelasticity. They proposed a special technique called "sandwich" technique, where the polariscope (polarizer and analyzer) were cemented in the interior of the considered body. They studied the stress fields in a sphere and a long thin-walled hole cylinder subjected to the steady-state thermal field obtained by dry ice.

F. Zandman, S.S. Redner, and D. Post [49] investigated the application of photoelastic coatings to the analysis of thermal fields. They indicated the following necessary conditions for coating to this purpose :

(1) Thermal conductivity of coating has to be equal to that of the workpiece.

(2) Coefficient of thermal expansion equal to that of the structure, or workpiece.

(3) Strain-optical sensitivity invariable with temperature.

Similarly to the method proposed by Vardanian and Prigorovski [46] T. Slot [50] proposed the photoelastic simulation of thermal stresses by mechanical prestraining, using frozen stress method. He analyzed disc composed of two semidiscs, one of them was prestrained by frozen stresses. He investigated also the model of steam generator of the U-tube type where the tube sheet was a thick, perforated plate, which was divided into an inlet side and an outlet side by a narrow, unperforated divider lane. The temperature difference was given between the two sides of the lane.

The prestraining method was applied also to the case of a composite structure built of materials with different coefficients of thermal expansion.

At all these cases the thermal stresses being simulated were produced by a one-dimensional thermal strain field. Because Poisson's ratio of photoelastic material is close to 1/2, when used in conjonction with the freezing technique, proper simulation of thermal stresses by mechanical prestraining is formally limited to this class of problems.

There is one more consequence of Poisson's ratio being 1/2, namely, that the prototype stresses predicted on the basis of the test results are formally correct only if Poisson's ratio of the prototype material is also 1/2. This is not usually the case (the common materials have $\nu \approx 0.3$). This difference induces new difficulties in final analysis applied to the stress field in the prototype.

A very interesting equipment to study of thermal stresses was proposed by C.P. Burger [51] who achieved the temperature differences using dry aviation kerosene as a heating and cooling liquid throughout the system. He had two reservoirs, one cold and the other hot. By a system of pumpes, it was possible to control the temperature regime. The model, a bolted-up assembly which represented part of a semi-infinite flat plate, made from Araldite B, and sandwiched between two circular polarizers. The temperature was measured by a system of thermocouples. A very exhaustive analysis of the existing stress field for different thermal conditions was worked out.

E. Hosp [52] presented a study giving a series of results at different models connected with problems in Civil Engineering (plates, dams on elastic foundations, rings, etc.), analyzing the whole process, including the analysis of similarity laws under different thermal conditions, all at steady-state temperature fields.

A series of papers discussing thermoelastic problems by means of Photoelasticity was worked out last ten years in connection with the design of nuclear power pressure vessels. These experiments were very often combined also with some other methods of experimental stress analysis (brittle coatings, strain gages, moiré method).

Chapter 4.
Application of Holography to Photoelasticity.

An exhausive presentation of Holography and Hologram Interferometry applied to Photoelasticity is given in Ref.[73]. Recently, M. E. Fourney and K.V. Mate published a paper (see Ref. 74) concerned to the explanation of holography phenomena as applied to photoelasticity. The main part of this very complete paper will be given in the following text.

4. 1. The Fourney-Mate's Analysis of Holography and Hologram Interferometry Applied to Photoelasticity.

M.F. Fourney and K.V. Mate[74] have given some additional contributions to the general explanation of phenomena occuring during the recording and the reconstruction process, discussing two-dimensional problems using the double exposure holograms. They mentioned that the reconstructed wave front form the unstressed model is need as a reference to which the reconstructed wavefront from the stressed model is compared. The result is expressed by isochromatic and isopachic patterns superimposed on the reconstructed image of the model.

The experimental arrangement (see Fig. 30) is like a standard circular polariscope where the reference acts as an analyzer. The $\lambda/4$ plates are oriented to give a light field when the polariscope is viewed through an analyzer with the same orientation as the polarization of the reference beam.

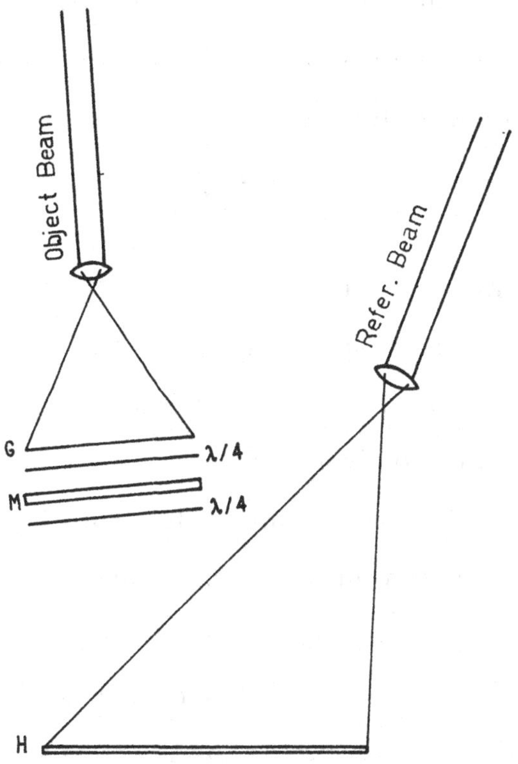

Fig. 30

4.1.1. The Analysis of the Double Exposed Hologram.

Let the reference-beam wave front striking the hologram plate be given by

$$\underset{\sim}{i} e^{i(\omega t + \vartheta)}$$

where ϑ is the spatial phase factor determined by the experiment-equipment geometry.

When there is no model in the circular polariscope, the wave-front from the ground glass point striking the holograph may be represented by

$$\underset{\sim}{i} e^{i(\omega t + \varphi)}$$

where φ depends on the geometry of the ground glass and the two quarter wave plates. The unit vector $\underset{\sim}{i}$ indicates that the reference beam wavefronts have the same polarization orientation.

However, when the model is placed in the light path the phase of the object beam will be changed by some value α (model is unstressed) and by α' and α'' for the stressed model in the directions parallel to the principal stresses σ_1 and σ_2 Thus, for the unstressed model the wave front is

$$\underset{\sim}{i} e^{i(\omega t + \varphi + \alpha)}$$

and for the stressed model

$$\underset{\sim}{i}\frac{1}{2}\left[e^{i(\omega t+\varphi+\alpha')} + e^{i(\omega t+\varphi+\alpha'')}\right]$$

(see Ref. [74]).

The $\underset{\sim}{j}$ component to this wavefront, which exists too, is here ignored, because it has been assumed that the polarization of the reference beam is in the $\underset{\sim}{i}$ direction.

The total wavefront striking the hologram plate during the first exposure is

$$\underset{\sim}{E}_1 = \left[e^{i(\omega t+\varphi+\alpha)} + e^{i(\omega t+\vartheta)}\right]\underset{\sim}{i} . \qquad (4.1)$$

The intensity at the film plate is given by the expression

$$I_1 = \underset{\sim}{E}_1\underset{\sim}{E}_1^* = 2 + e^{i(\varphi+\alpha-\vartheta)} + e^{-i(\varphi+\alpha-\vartheta)} . \qquad (4.2)$$

The wavefront to the second exposure, when the model is loaded :

$$\underset{\sim}{E}_2 = \left[e^{i(\omega t+\vartheta)} + \frac{1}{2}e^{i(\omega t+\varphi+\alpha')} + \frac{1}{2}e^{i(\omega t+\varphi+\alpha'')}\right]\underset{\sim}{i} . \qquad (4.3)$$

The intensity pattern at the film during the second exposure : (see page 178)

$$I_2 = \frac{3}{2} + \frac{1}{2}e^{i(\varphi+\alpha'-\vartheta)} + \frac{1}{2}e^{-i(\varphi+\alpha'-\vartheta)} +$$

(4.4)
$$+ \frac{1}{2}e^{i(\varphi+\alpha''-\vartheta)} + \frac{1}{2}e^{-i(\varphi+\alpha''-\vartheta)} +$$

$$+ \frac{1}{4}e^{i(\alpha'-\alpha'')} + \frac{1}{4}e^{-i(\alpha'-\alpha'')} .$$

After processing, the interference pattern recorded on film, i.e. the hologram, acts as a transmission difraction grating. The transmission factor T of the difraction grating depends on the total exposure. Since there are two exposures, T is proportional to the sum of I_1 and I_2 :

(4.5)
$$T = K[I_1 + I_2] .$$

In the reconstruction process, the holograph is returned to its original location and illuminated by the reference beam. The grating pattern on the holograph attenuates the reference beam as it passes through the holograph. The amplitude of the transmitted wave front emerging from the backside of the holograph is given by the product of the incident wavefront and the transmission factor of holograph :

$$\underset{\sim}{E}_T = Te^{i(\omega t + \delta)}\underset{\sim}{i} . \tag{4.6}$$

By substituting Eqs. (4.2), (4.4) and (4.5) into (4.6) we get :

$$\begin{aligned} \underset{\sim}{E}_T &= K[I_1 + I_2]e^{i(\omega t+\delta)}\underset{\sim}{i} = \\ &= K_1 e^{i(\omega t+\delta)}\underset{\sim}{i} + \\ &+ K_2 e^{i(\omega t+\varphi)}\left[e^{i\alpha} + \frac{1}{2}e^{i\alpha'} + \frac{1}{2}e^{i\alpha''}\right]\underset{\sim}{i} + \\ &+ K_3 e^{i(\omega t-\varphi+2\delta)}\left[e^{-i\alpha} + \frac{1}{2}e^{-i\alpha'} + \frac{1}{2}e^{i\alpha''}\right]\underset{\sim}{i} + \\ &+ K_4 \frac{1}{4}e^{i(\omega t+\delta+\alpha'-\alpha'')}\underset{\sim}{i} + \\ &+ K_5 \frac{1}{4}e^{i(\omega t+\delta-\alpha'+\alpha'')}\underset{\sim}{i} . \end{aligned} \tag{4.7}$$

The first term of the wavefront emerging from the holograph is the undiffracted reference beam. The second term represents the wave front of the virtual image. The third term represents the real image. The forth and the fifth terms represent the wavefronts diffracted by a transmission pattern on the holograph similar to the isochromatic pattern seen on the stressed model. These terms are noted only to acknoledge their existence which may be observed when the variations of the equipment geometry are used, for instance instead of ground glass to apply one point light source. These terms have no effect on the reconstructed virtual image.

The virtual image is represented by the second set of terms :

$$E_v = e^{i(\omega t + \varphi)}\left[e^{i\alpha} + \frac{1}{2}e^{i\alpha'} + \frac{1}{2}e^{i\alpha''}\right]i \qquad (4.8)$$

where the constant K_2 is choosen as unity for convenience.

The intensity pattern of this image is then :

$$I = E_v E_v^* = \frac{3}{2} + \cos(\alpha - \alpha') + + \cos(\alpha - \alpha'') + \frac{1}{2}\cos(\alpha' - \alpha'') . \qquad (4.9)$$

Now, the relationship between the principal stresses and the isochromatics and isopachics on the reconstructed image will be taken into account.

The phase changes due to the model are:

$$\alpha = \frac{2\pi}{\lambda} n_0 t \qquad (4.10a)$$

$$\alpha' = \frac{2\pi}{\lambda}\left[n_1 t + n(t - t')\right]$$

$$\alpha'' = \frac{2\pi}{\lambda}\left[n_2 t' + n(t - t')\right] \tag{4.10b}$$

where

n is index of refraction of the air,

n_0 is index of refraction of the unstressed model,

n_1, n_2 indices of refraction of the stressed model in the directions of the principal stresses,

t, t' initial and final thickness of the model.

The Maxwell-Neumann stress-optic law :

$$\begin{aligned} n_1 - n_0 &= A\sigma_1 + B\sigma_2 \\ n_2 - n_0 &= B\sigma_1 + A\sigma_2 . \end{aligned} \tag{4.11}$$

The final thickness of the model for the case of plane stress :

$$t' = t - \frac{\nu}{E}(\sigma_1 + \sigma_2)t . \tag{4.12}$$

Substituting last three equations into Eq. (4.9) we obtain after neglecting small-order terms :

$$\begin{aligned} I = 1 &+ 2\cos\frac{\pi t}{\lambda}C(\sigma_1 - \sigma_2)\cos\frac{\pi t}{\lambda}(A' + B')(\sigma_1 + \sigma_2) + \\ &+ \cos^2\frac{\pi t}{\lambda}C(\sigma_1 - \sigma_2) \end{aligned} \tag{4.13}$$

where

$$A' = A - \frac{\nu}{E}(n_0 - n)\,; \quad B' = B - \frac{\nu}{E}(n_0 - n) \tag{4.14}$$
$$C = A' - B' = A - B\,.$$

The intensity pattern depends on both the sum and the difference of the principal stresses, i.e. the isochromatic pattern is modulated by the isopachic pattern.

The loci of points where

$$\frac{\pi}{\lambda}C(\sigma_1 - \sigma_2)t = m\pi \qquad (m = \text{integer}) \tag{4.15}$$

usually correspond to a light isochromatic fringe in a standard light-field pattern. Here, however, the light isochromatic is modulated by the isopachic pattern giving

$$I = 2 + 2\cos\frac{\pi}{\lambda}(A' + B')(\sigma_1 + \sigma_2)t\,. \tag{4.16}$$

At points where

$$\frac{\pi}{\lambda}C(\sigma_1 - \sigma_2)t = \frac{2m+1}{2}\pi \tag{4.17}$$

usually corresponds to a dark isochromatic pattern ; the intensity is :

$$I = 1\,. \tag{4.18}$$

Thus, the fringes corresponding to the dark isochromatics are the half-tone gray fringes curving through the isopachics. Furthermore, as isopachics intersect the isochromatic lines, their intensities revers from light to dark and vice versa.

Thus, the entire intensity pattern consists of two independent families of fringe, $(\sigma_1 + \sigma_2)$ and $(\sigma_1 - \sigma_2)$, and the plane problem can be solved directly. If N_c and N_p are the fringe orders of an isochromatic fringe and an isopachic fringe, respectively, we have :

$$\sigma_1 + \sigma_2 = \frac{\lambda N_p}{(A' + B')t} ; \quad |\sigma_1 - \sigma_2| = \frac{\lambda N_c}{Ct} . \qquad (4.19)$$

The result is similar to that of Nisida-Saito[18].

The last term in Eq. (4.13) is the intensity pattern of the isochromatic fringes seen if the second exposure only is reconstructed. This pattern is superimposed over the identical isochromatic fringe pattern formed from the interference of the first and second exposure wavefronts.

Fourney and Mate verified their theory at some examples (a circular ring compressed diametrally, a deep beam with a central concentrated load).

4. 2. Recording of Isoclinics by Holography. [74]

Fourney and Mate proposed one more theory of recording and reconstructing the polarization of light emitted by a photoelastic model. By this method, it is possible, by viewing the hologram, to determine the entire family of isoclinics a posteriori. To this purpose the state of polarization of the photoelastic model has to be recorded. This technique is to the idea proposed by Lohmann [75] and Bryndal [76] by the use of two orthogonally polarized reference beams (Fig. 31).

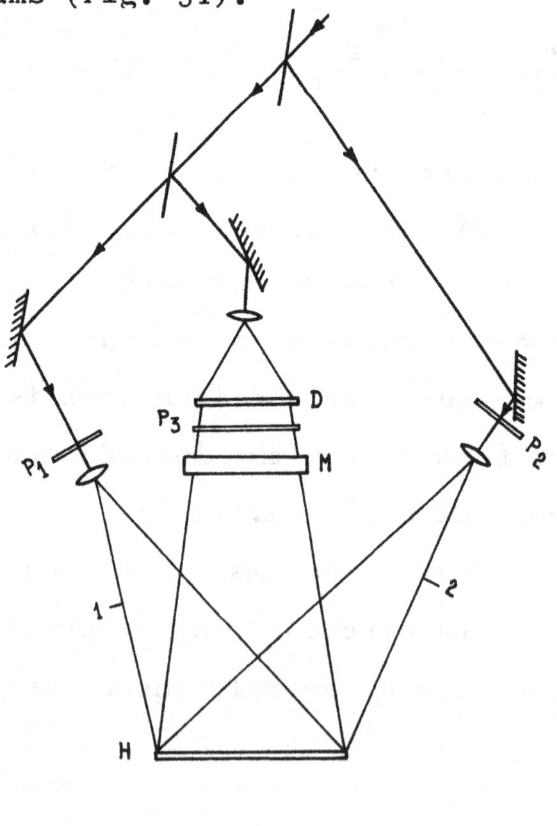

Fig. 31

Fourney and his colaborators showed that both the light- and dark-field isochromatic patterns can be obtained by this technique using only one hologram.

Two interference patterns are now superimposed and recorded simultaneously by the hologram.

If the hologram is illuminated by the original reference beams, the reference wavefronts are diffracted by the hologram and combined vectorially to produce the polarization of light emitted by the object. However, it is necessary to maintain the location and the polarization of the reference beams with respect to each other and with respect to the holograph from formation to reconstruction. To reconstruct the object's polarization, the phase relationship has to be maintained.

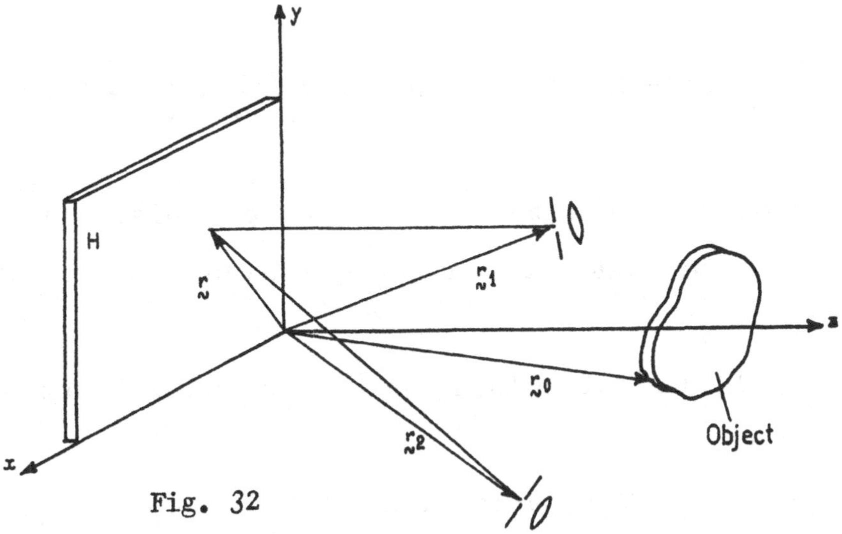

Fig. 32

Let xy-plane (see Fig. 32) of the coordinate system be in the plane of hologram, and the position of a point on the holograph be defined by $\underset{\sim}{r}$. The same for the position vectors $\underset{\sim}{r}_0$, $\underset{\sim}{r}_1$, $\underset{\sim}{r}_2$.

It is dealt with one dipole radiator and assumed if its wave fronts, including polarization, can be reconstructed that of the entire object will be reconstructed.

The wavefront from a model point or ground glass point falling on the hologram plate can be represented by

$$\underset{\sim}{E}_0 = e^{i(\omega t+\varphi)}\underset{\sim}{i} + e^{i(\omega t+\varphi_\perp)}\underset{\sim}{j}$$

where φ and $\varphi_\perp$ are the spatial phase factors given by

$$\varphi = \frac{2\pi}{\lambda}|\underset{\sim}{r} - \underset{\sim}{r}_0| ; \quad \varphi_\perp = \frac{2\pi}{\lambda}|\underset{\sim}{r} - \underset{\sim}{r}_0| + D .$$

The constant D determines the type of polarization : if $D = 0$, the model is emitting a plane polarized wave front ;

if $D = \pi/2$, the circularly polarized wave front will be emitted.

The wavefronts striking the hologram

plate from the reference beams can be given by :

$$\underset{\sim}{E}_1 = e^{i(\omega t + \vartheta)} \underset{\sim}{i} \; ; \quad \underset{\sim}{E}_2 = e^{i(\omega t + \vartheta_\perp)} \underset{\sim}{j}$$

where

$$\vartheta = \frac{2\pi}{\lambda} |\underset{\sim}{r} - \underset{\sim}{r}_1| \; ; \quad \vartheta_\perp = \frac{2\pi}{\lambda} |\underset{\sim}{r} - \underset{\sim}{r}_2| \, .$$

The total amplitude at point $\underset{\sim}{r}$ on the hologram plate is given by the sum :

$$\begin{aligned} \underset{\sim}{E} &= \underset{\sim}{E}_0 + \underset{\sim}{E}_1 + \underset{\sim}{E}_2 = \\ &= e^{i(\omega t + \varphi)} \underset{\sim}{i} + e^{i(\omega t + \varphi_\perp)} \underset{\sim}{j} + e^{i(\omega t + \vartheta)} \underset{\sim}{i} + e^{i(\omega t + \vartheta_\perp)} \underset{\sim}{j} \end{aligned} \tag{4.20}$$

Each orthogonal component of the object's wave front interfers with its respectively polarized reference beams forming two patterns at the hologram plate :

$$\begin{aligned} I_i &= 2 + e^{i(\varphi - \vartheta)} + e^{-i(\varphi - \vartheta)} \\ I_j &= 2 + e^{i(\varphi_\perp - \vartheta_\perp)} + e^{-i(\varphi_\perp - \vartheta_\perp)} \end{aligned} \tag{4.21}$$

The total intensity pattern recorded by the plate :

$$I = I_i + I_j \, .$$

The transmission factor T of the hologram is then

given by the expression :

$$(4.22)\ T = \text{const}\left[4 + e^{i(\varphi-\vartheta)} + e^{-i(\varphi-\vartheta)} + e^{i(\varphi_{\perp}+\vartheta_{\perp})} + e^{-i(\varphi_{\perp}-\vartheta_{\perp})}\right] .$$

To determine the restrictions on the reconstruction beams in order to reproduce the polarization of light emitted by the object, let the positions of the reconstruction beam sources be represented by $\underset{\sim}{r}_1'$ and $\underset{\sim}{r}_2'$ with wavefronts given by

$$\underset{\sim}{E}_1' = e^{i(\omega t+\vartheta')}\underset{\sim}{i} \ ; \quad \underset{\sim}{E}_2' = e^{i(\omega t+\vartheta_{\perp}')}\underset{\sim}{j}$$

where

$$\vartheta' = \frac{2\pi}{\lambda}|\underset{\sim}{r} - \underset{\sim}{r}_1'| \ ; \quad \vartheta_{\perp}' = \frac{2\pi}{\lambda}|\underset{\sim}{r} - \underset{\sim}{r}| \ .$$

The incident wavefront passing through the holograph is :

$$\underset{\sim}{E}_{inc} = \underset{\sim}{E}_1' + \underset{\sim}{E}_2' \ .$$

The wavefront emerging from the holograph is then :

$$\underset{\sim}{E}_T = T[\underset{\sim}{E}_1' + \underset{\sim}{E}_2'] =$$

$$= \text{const}\left[e^{i(\omega t+\vartheta')}\underset{\sim}{i} + e^{i(\omega t+\vartheta_{\perp}')}\underset{\sim}{j}\right] +$$

$$+ \text{const}\left[e^{i(\omega t+\varphi-\vartheta+\vartheta')} + e^{i(\omega t-\varphi+\vartheta+\vartheta')} + \right.$$

$$\left. + e^{i(\omega t+\varphi_\perp-\vartheta_\perp+\vartheta')} + e^{i(\omega t-\varphi_\perp+\vartheta_\perp+\vartheta'_\perp)}\right]\underset{\sim}{i} + \qquad (4.23)$$

$$+ \text{const}\left[e^{i(\omega t+\varphi_\perp-\vartheta+\vartheta'_\perp)} + e^{i(\omega t-\varphi_\perp+\vartheta_\perp+\vartheta'_\perp)} + \right.$$

$$\left. + e^{i(\omega t+\varphi-\vartheta+\vartheta'_\perp)} + e^{i(\omega t-\varphi+\vartheta+\vartheta'_\perp)}\right]\underset{\sim}{j} .$$

The first row represents the part of the reconstruction wavefront that passes through the holograph without being diffracted. This part contains no informations about the object.

There are eight individual wavefronts that are diffracted by the holograph that do carry informations about the object :

There are two virtual-image wavefronts having orthogonal polarizations represented by the first term in each component.

There are two real-image wavefronts represented by the second term.

The third and fourth terms in each component represent the virtual and real crosstalk images, respectively. These images are due to the diffraction of one component of polarization of the reconstruction wavefront by the interference pattern formed by the orthogonal component of the reference wavefront.

If the reconstructed virtual wavefronts are to add vectorially to yield the object's original polarization, the phase relationship between the reconstructed components must be the same as that between the components of the original object wavefront. It can be shown that the necessary condition is

$$\underset{\sim}{r}_1 = \underset{\sim}{r}_1' ; \quad \underset{\sim}{r}_2 = \underset{\sim}{r}_2' ,$$

i.e. the reconstruction beam sources must be located at the same position as the reference beam sources. Thus, to reconstruct the polarization of the light emitted by the object, it is necessary to maintain the polarization and location of the reference beams with respect to the holograph from the formation process through the reconstruction process. In order to obtain it, it is necessary the hologram be developed in place, clamping it to a rigid base plate [74].

In this case, the polarization is recorded. The isoclinic fringe is seen as a gray fringe that intersects the isochromatic fringes. This isoclinic fringe pattern results from using circularly polarized light to illuminate the model and viewing through a linear polarizer.

If the model is illuminatee by circularly polarized light the wave front emerging from the object can be represented by

$$\underset{\sim}{E}_0 = e^{i(\omega t+\alpha')}\underset{\sim}{i} + e^{i(\omega t+\alpha''+\frac{\pi}{2})}\underset{\sim}{j} \qquad (4.24)$$

where $\underset{\sim}{i}$ and $\underset{\sim}{j}$ are the directions of principal axes and α' and α'' are the optical phase retardations. If the analyzer is oriented at an angle β to one of the principal axes of the model, the wavefront emerging from the analyzer is :

$$\underset{\sim}{E}_\alpha = \left[\cos\beta e^{i(\varphi+\alpha')} + \sin\beta e^{i(\varphi+\alpha''+\frac{\pi}{2})}\right]\underset{\sim}{i} \,. \qquad (4.25)$$

The intensity pattern is then :

$$I = \underset{\sim}{E}_\alpha \underset{\sim}{E}_\alpha^* = 1 + \sin 2\beta \sin(\alpha' - \alpha'') \,. \qquad (4.26)$$

When $\beta = 0$, i.e. when the axis of analyzer is the same as the principal axis, then $I = 1$ or 1/2 (maximum). Thus, the isoclinics for this particular case are gray or half-tone fringes. Note also that $\sin(\alpha' - \alpha'')$, which determines the isochromatic pattern, is modulated by the factor $\sin 2\beta$. It follows that the isochromatic fringe changes sense as it crosses the isoclinic. This effect has been verified experimentally.

Lately, a series of papers was published by different Authors, as C. A. Sciamarella, G. D. Chirico and T. Y. Chang [77]; by J.D. Hovanesian[78]-[80], and many others, treating different applications of Holography in displacement and strain analysis, thermoelasticity, dynamical problems, theory of crack propagation, combination of two and three reconstructed waves in Photoelasticity, etc.

The domain of Holography, due to recent developments in laser technique, Coherent Optics in generally, than of the computer and electronic devices, has been still in a rapidly developmental stage.

References.

[1] E.G. Coker, L.N.G. Filon : "A Treatise on Photoelasticity" ; Cambridge, At the University Press, 1957.

[2] H.T. Jessop : "Photoelasticity" ; Handbuch der Physik, Band VI, Springer, 1958.

[3] J.W. Dally, W.F. Riley : "Experimental Stress Analysis"; Mc Graw-Hill, New York, 1965.

[4] E.H. Dill : "On the Theory of Photoviscoelasticity" Univ. of Washington, Dept. Aeronautics and Astronautics, Report 63-1, January, 1963.

[5] J.T. Pindera : "Response of Photoelastic Systems"; Solid Mech. Div., Univ. of Waterloo, Ontario, Canada, Report 17, August, 1969.

[6] H. Wolf : "Spannungsoptik" ; Springer, 1961.

[7] J.A. Straton : "Electromagnetic Theory" ; Mc Graw Hill, 1941.

[8] C. Truesdell, R. Toupin : "The Classical Field Theories" ; Enc. of Ph., V III/1, Springer, 1960.

[9] R.D. Mindlin, L.E. Goodman : "The Optical Equations of Three-dimensional Photoelasticity" ; Jnl. of Appl. Ph., 20,(1), 89-97, (1949).

[10] R.C. O'Rourke :"Three-dimensional Photoelasticity"; Jnl. Appl. Ph., 22, (7), 872-878, (1951).

[11] D.C. Drucker, W.B. Woodward : "Interpretation of Photoelastic Transmission Patterns for a Three-dimensional Model", Jnl. Appl.. Ph., (1954).

[12] H.K. Aben : "Optical Phenomena in Photoelastic Models by the Rotation of Principal Axes", Exp. Mech., 6 (1), 13-22, (1966).

[13] H.K. Aben : " On the Application of Photoelastic Coatings by the Investigation of Shells", Izv. Akad. Nauk SSSR, Mekh. i Mashinostr., 7 (6), (1964).

[14] H.K. Aben: "Optical Theory of the Multilayer-reflection Technique for Three-dimensional Photoelastic Studies", Exp. Mech., 9 (1), 25-30, (1969).

[15] F. Neumann :"Die Gesetze der Doppelbrechung des Lichtes in komprimierten oder ungleichförmig erwärmten unkrystallinischen Körpern", Abh. d. Kön. Akad. d. Wissensch. zu Berlin, Pt.II, (1841).

[16] V.L. Ginsburg: "On the Investigation of Stress by the Optical Method" (in Russian), Zb. Tekh. Fiz., 14 (3), (1944).

[17] D. Post: "The Generic Method of the Absolute-retardation Method of Photoelasticity", Exp. Mech., 7 (6), 233-241, (1967).

[18] M. Nisida, H. Saito : "A New Interferometric Method of Two-dimensional Stress Analysis", Exp. Mech., 4 (12), 366-376, (1964).

[19] E. Mönch:"Similarity and Model laws in Photoelastic Experiments", Exp. Mech., 4 (5). 141-150, (1964).

[20] R.D. Mindlin :"A Mathematical Theory of Photo-Viscoelasticity" Jnl. Of Appl. Ph., 20 206-216,(1949).

[21] W.T. Read : "Stress Analysis for Compressible Viscoelastic Materials", Jnl. of Appl. Ph., 21 (7), 671-674, (1950).

[22] R.J. Arenz : "Theoretical and Experimental Studies of Wave Propagation in Visco-elastic Materials", Disertation, Calif. Inst. Techn., June, (1964).

[23] M.L. Williams, R.J. Arenz :"The Engineering Analysis of Linear Photoviscoelastic Materials", Exp. Mech., 4 (9), 244-(1964).

[24] P.S. Theocaris, D. Mylonas: "Viscoelastic Effects in Birefringente Coatings", Jnl. Appl. Mech., 28, Trans ASME, 83, 601-607, Dec., (1961).

[25] E.H. Dill: " On the Theory of Photoviscoelasti-city", Univ. Wash. dept. Aeron. and Astronautics; Report 63-1, January, (1963).

[26] I.M. Daniel: " Quasi-static Properties of a Photoviscoelastic Material", Exp. Mech., 5 (3), 83-89, (1965).

[27] I.M. Daniel: " Dynamic Properties of a Photo-viscoelastic Material", Exp. Mech., 6 (5), 225-234, (1966).

[28] J.T. Pindera: " Remarks on Properties of Photo-viscoelastic Material", Exp. Mech., 6 375-380, (1966).

[29] V. Brčić, M. Nešović: " A Contribution to the Photoviscoelastic Investigation of

Structures", Materijali i Konstr., 10 (2), 3-14, (1966).

[30] V. Brčić, M. Nešović: "Application of a Low-modulus Material at Photoelastic Investigation of the Djerdap Dam", Proc. 7th Congr. of Yug. Comittee for High Dams, 193-197, Sarajevo, (1966).

[31] M.G. Sharma, C.K. Lim :"Experimental Investigation on Fracture of Viscoelastic Materials under Biaxial-stress field", Exp. Mech., 8 (5), 202-209, (1968).

[32] R.M. Hackett, E.M. Krokosky :"A Photoviscoelastic Analysis of Time-dependent Stresses in a Polyphase System", Exp. Mech., 8 (12), 539-547; (1968).

[33] H.F. Brinson :"Mechanical and Optical Viscoelastic Characterization of Hysol 4290", Exp. Mech., 8 (12), 561-566, (1968).

[34] M.L. Williams :"Structural Analysis of Viscoelastic Materials", AIAA Journal, Vol. 2, May, (1964).

[35] R.A. Shapery :"Approximate Methods of Transform Inversion for Viscoelastic Stress Analysis", Proc. Fourth U.S. Natl. Congr., Appl. Mech., 2, (1962).

[36] R.L. Adkins :"Design Considerations and Analysis of a Complex-modulus Apparatus", Exp. Mech. , 6 (7), 362-367, (1966).

[37] A.S. Miguel, R.H. Silver :"A Normal-incidence Reflective Polariscope for Viscoelastic Measurements", Exp. Mech., 5 (10), 345-362, (1965).

[38] P.S. Theocaris :"A Review of Rheo-Optical Properties of Linear High Polymers", Exp. Mech., 5 (4), 105-114, (1965).

[39] T. Alfrey, P. Doty :"The Methods of Specifying the Properties of Viscoelastic Materials", Jnl. of Appl. Ph., 16,(1945).

[40] E.E. Weibel :"Thermal Stresses in Cylinders by the Photoelastic Method", (1938).

[41] G. Gerard, A.C. Gilbert :"Note on Photothermoelasticity", J. Aero. Sci., Vol. 33, July (1965).

[42] G. Gerard, A.C. Gilbert :"Photothermoelasticity: An Exploratory Study", Jnl. Appl. Mech., 24 355-360, September, (1957).

[43] H. Trampoch, G. Gerard :"An Exploratory Study of Three-Dimensional Photothermoelasticity", Jnl. Appl. Mech., 28 (3),35-40, (1961).

[44] G. Gerard :"Progress in Photothermoelasticity", Photoelasticity, Proc. of the Int. Symp., Chicago, (1961), Editor M. M. Frocht.

[45] J.D. Hovanesian, H.C. Kowalski :"Similarity in Thermoelasticity", Exp. Mech., 7 (2), 82-84, (1967).

[46] G.S. Vardanian, N.L. Prigorovski:" Modelirovanie termouprugih naprazenii v polarizaciono opticeskom metode", Izv. Akad. Nauk SSSR, Mech. i Masstr., Moscow, (1962).

[47] V. Švec :"Sledovani tepelnych napeti pomoci fotoelasicimetrie" Strojirenstvi, 13 (3), 208-214, Praha, (1963).

[48] Z. Orlós, Z. Dylag :"Über spannungsoptische Untersuchung von Wärmespannungen (Int. Spann. opt. Symp., Deutsche Akad. der Wissensch., Berlin, (1962).

[49] F. Zandman, S.S. Redner, D. Post :"Photoelastic-coating Analysis in Thermal Fields", Exp. Mech., 3 (9), 215-221, (1963).

[50] T. Slot :"Photoelastic Simulation of Thermal Stresses by Mechanical Prestraining", Exp. Mech., 5 (9), 273-282, (1965).

[51] C.P. Burger :"A Generalized Method for Photoelastic Studies of Transient Thermal Stresses", Exp. Mech., 9 (12), 529-537, (1969).

[52] E. Hosp :"Experimentelle Bestimmung von Wärmespannungen in Bauteilen auf spannungsoptischem Wege", Die Bautechnik, November, (1960).

[53] D. Gabor :"A New Microscopic Principle", Nature, 161, 777, (1948).

[54] E. Leith, L. Upatknies : Jnl. Opt. Soc. Am., 52, 1123, (1963).

[55] E.N. Leith, A. Kozma, J. Upatknies, J. Marks, N. Massey :"Holographic Data Storage in Three-Dimensional Media", Appl. Optics, Vol. 5, 1303, August, (1966).

[56] G.W. Stroke :"An Introduction to Coherent Optics and Holography", Acad. Press, New-York-London, (1966).

[57] Mme Pauthier-Camier :"L'Holographie", Rev. Franc. de Mec., 26, 89-95, (1968).

[58] R.L. Powell, K.A. Stetson :"Interferometric Vibra-

tion Analysis by Wavefront Reconstruction", J; Opt. Soc. Am., 55 (12), 1593, (1965).

[59] K.A. Stetson, R.L. Powell :"Interferometric Hologram Evaluation and Real-Time Vibration Analysis of Diffuse Objects", J. Opt. Soc. AM;, 56 (12), 1694-1695, (1965).

[60] K.A. Stetson, R.L. Powell :"Hologram Interferometric ", J. Opt. Soc. Am., 56 (9), 1161-1166, (1966).

[61] B.P. Hildebrand, K.A. Haines :"Interferometric Measurements Using the Wavefront Reconstruction Technique", Appl. Opt., 5 (1), 172-173, (1966).

[62] K.A. Haines, P.B. Hildebrand :"Surface Deformation Measurement Using the Wavefront Reconstruction Technique", Appl. Opt. 5 (4), 595-602, (1966).

[63] M.E. Fourney :"Application of Holography to Photo elasticity", Exp. Mech., 9 (1), 33-38.

[64] J.D. Hovanesian, V. Brčić, R.L. Powell : "A New Stress-Optic Method : Stress-Holo-Interferometry", Exp. Mech., 8 (8), 362-368, (1968).

[65] V. Brčić :" Hologram Interferometry and its application to Experimental Stress Analysis", Transactions, Inst. Jar. Cerni, Belgrade, 43, 26-30, (1967).

[66] R.L. Powell, J.D. Hovanesian, V. Brcic :"Hologram Interferometry with Birefringent Objects", (unpublished text), Ann Arbor, 1968.

[67] D. Gabor :"Microscopy by Reconstructed wavefronts", Proc. Roy. Soc., London, A 197, 454-489, (1949).

[68] E.N. Leith, J. Upatknies :"Reconstructed Wavefronts and Communication Theory", J. Opt. Soc. Am., 52 (10).

[69] R.E. Brooks, L.O. Hoflinger, R.F. Wuerker, R.A. Briones : J. Appl. Ph.,37, 642,(1966).

[70] C.A. Sciamarella :"Moiré-fringe Multiplication by Means of Filtering and a Wave-front Reconstruction Process", Exp. Mech., 9 (4), 179-185, (1969).

[71] W.A. Gottenberg :"Some Applications of Holographic Interferometry", Exp. Mech., 8 (9), 405-410, (1968).

[72] E.R. Robertson, J.M. Harvey :"The Engineering Uses of Holography", Cambridge University Press (to be published).

[73] V. Brčić :"Application of Holography and Hologram Interferometry to Photoelasticity", CISM, Udine, (1970).

[74] M.E. Fourney, K.V. Mate :"Further Applications of Holography to Photoelasticity", Exp. Mech., 10 (5), 177-186, (1970).

[75] A.W. Lohmann : Appl. Opt., 4, 1667, (1965).

[76] O. Bryngdal: Jnl. Opt. Soc. Am., 57, 545, (1967).

[77] C.A. Sciamarella, G. Di Chirico, T.Y. Chang : "Moiré-Holographic Technique for Three-Dimensional Stress Analysis", J. of Appl. Mech., March, (1970).

[78] J.D. Hovanesian :"New Applications of Holography

to Thermoelastic Studies", Fourth Int. Conf. on Stress Analysis, Cambridge, England, May, (1970).

[79] J.D. Hovanesian :"Interference of Two and Three Reconstructed Waves in Photoelasticity", U.S. Navy Jnl. of Underw. Acoustics, October, (1968).

[80] J.D. Hovanesian, E.C. Zobel :"Application of Photoelasticity in the Study of Crack Propagation", Detroit, 1969.

[81] J.R. Nicolas :"Contributions à la détermination de la sommes des contraintes principales au moyen de l'Holographie", Rev. Franc. de Mec., 28, 48-56, (1968).

Contents.

INTERNATIONAL CENTRE FOR MECHANICAL SCIENCES

A. AJOVALASIT
UNIVERSITY OF PALERMO

EXPERIMENTAL METHODS IN PHOTOTHERMOELASTICITY

UDINE 1970

COURSES AND LECTURES

F O R E W O R D

This lecture contains a short survey of the experimental methods employed in the photoelastic analysis of thermal stresses.

I am grateful to the Secretary General Prof. Luigi Sobrero and to Prof. Vlatko Brčić for inviting me to deliver this lecture at CISM in Udine.

Introduction

Thermal stresses can be analyzed by the photoelastic method in two different ways :

1) Photoelastic simulation of thermal stresses by using the prestraining methods [1] , [2] or the Biot's analogy [3] , [4], [5]

2) Photothermoelasticity by producing in the model the required temperature distribution [6], [7], [8], [9].

In this lecture some typical features of photothermoelasticity are treated. For a general view of the subject the reader must refer to the chapter 3 of the course on "Photoelasticity in Theory and Practice" delivered by Professor V. Brčić.

The photothermoelastic method differs from the classic photoelasticity in the loading system which is thermal rather than mechanical. Then, in order to perform a photothermoelastic investigation, it is necessary:

1) to produce and measure the required temperature distribution in the model;

2) to calibrate the photoelastic material;

3) to record the photoelastic data;

4) to separate the principal stresses.

Thermal Loading of Models

Most photoelastic materials have physical properties which change quickly as the temperature rises above room temperature.

For this reason thermal gradients are obtained by refrigeration using mainly dry-ice or a mixture of ethylic alcohol and dry-ice [10], [11], [12], [13], [14], [15].

In some cases however a moderate heating of the model may be applied [16], [17], [18], [19].

Thermocouples embedded in the model are generally employed to measure the temperature distribution.

Fig. 1 shows an apparatus for the thermal loading of discs with axially symmetric temperature distribution.

Fig. 1. Thermal loading device (by courtesy of The Institution of Mechanical Engineers from ref. 15).

Two dry-ice cylinders supported by the plastic discs L are pressed on the model by means of the springs M . Two Plexiglas envelopes N prevent the gaseous carbon dioxide from touching the model.

Fig. 2a shows a light-field isochromatic pattern of a disc loaded with this apparatus. This disc was employed to study the thermal stresses that arise in discs with eccentric holes such as are employed in the rotors of multi-stage gas turbines. In particular the optimum spacing of holes was studied , in order to minimize the thermal stresses.

In fig. 2b an enlargement of one of the eccentric holes is shown (four singular points S_1, S_2, S_3 and S_4 are marked).

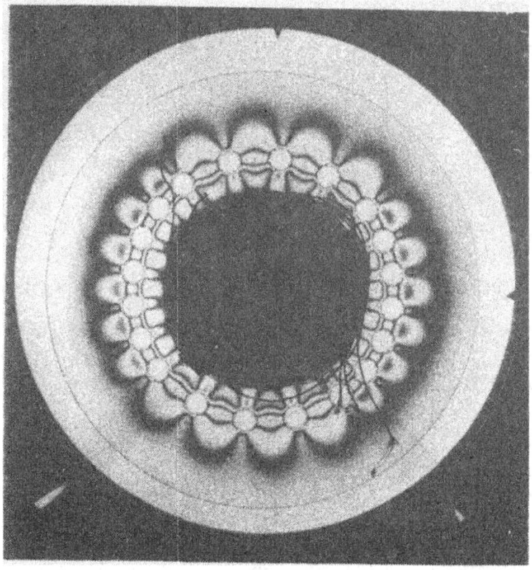

Fig. 2 a.

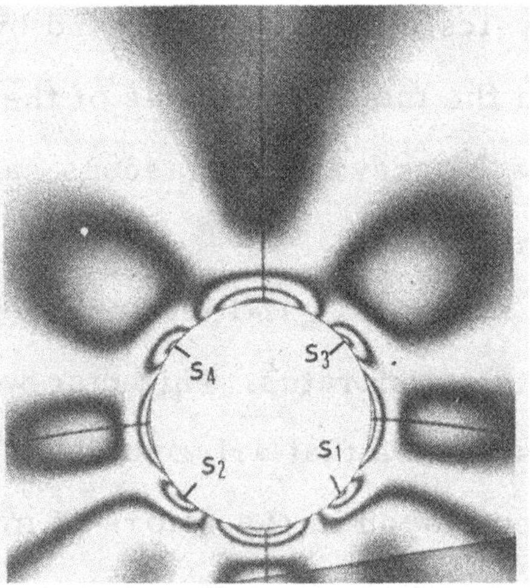

Fig. 2 b. Isochromatic pattern for a disc with eccentric holes under a thermal gradient. (by courtesy of the Institution of Mechanical Engineers from ref. 15).

Fig. 3 shows the dark-field isochromatic patterns in a turbine disc with blades using the same loading apparatus of Fig. 1.

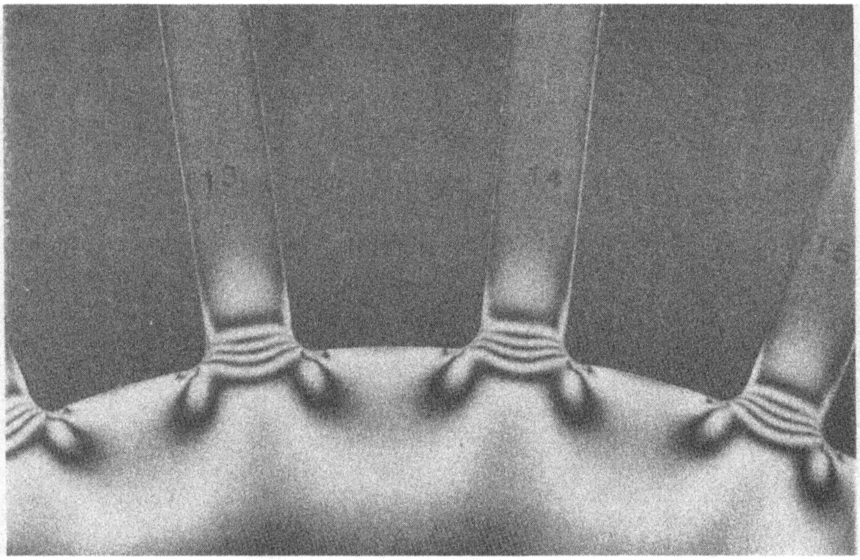

Fig. 3 a.

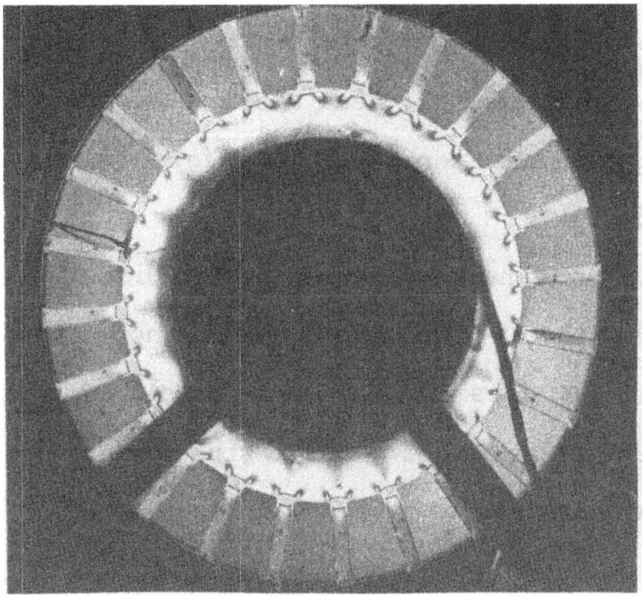

Fig. 3 b. Isochromatic pattern for a disc with blades under a thermal gradient.

Material Calibration

In the photothermoelastic investigations the results are usually expressed by the nondimensional stress indices :

$$K = \sigma/E\alpha\theta_1 \tag{1}$$

for plane stress systems, and

$$K' = (1 - \nu)\sigma/E\alpha\theta_1 \tag{2}$$

for plane strain systems,where

E is the Young's modulus

ν is the Poisson's coefficient

α is the coefficient of linear thermal expansion

θ_1 is the maximum temperature difference existing in the model.

The advantage of using these indices lies in the circumstance that they are invariant for the model and prototype if the product $E\alpha$ does not vary with the temperature [20] .

The relation between the stress index K (plane stress) and the fringe at a free boundary ($\sigma = 2fn/h$) is :

(3) $$K = \frac{\sigma}{E\alpha\theta_1} = 2n/Q_\alpha\theta_1 h$$

where

$Q_\alpha = E\alpha/F$ is the photothermoelastic figure of merit

f is the fringe value of material in shear

h is the thickness of model

The photothermoelastic figure of merit Q_α can be determined :

1) by a separate measurement of E , α and f ,

2) by a direct calibration method using a disc in which a known thermal gradient is produced.

If the disc has a central hole, the thermal stresses σ_r and σ_θ are given by the following equations :

$$\frac{\sigma_r}{E\alpha\theta} = \frac{1}{R^2}\left[\frac{R^2 - R_i^2}{R_e^2 - R_i^2}A(R_e) - A(R)\right]$$

$$\frac{\sigma_\theta}{E\alpha\theta_1} = \frac{1}{R^2}\left[\frac{R^2 + R_i^2}{R_e^2 - R_i^2}A(R_e) + A(R) - \frac{\theta}{\theta_1}R^2\right]$$

in which

R is the generic radius

R_i is the internal radius

R_e is the external radius

and

$$A(R) = \int_{R_i}^{R} \frac{\theta}{\theta_1} R\,dR$$

is calculated numerically from the temperature curve which is measured by means of thermocouples.

Substituting in equation (3) the values of $\sigma_\theta / E\alpha\theta_1$ and of n on the external boundary, the photothermoelastic figure of merit is determined. This calibration method was adopted in ref. 15 using the models before machining the eccentric holes.

Using materials with a high figure of merit a sufficient number of isochromatic fringes develops with low temperature difference, so that the variations of the physical properties of the material are minimized.

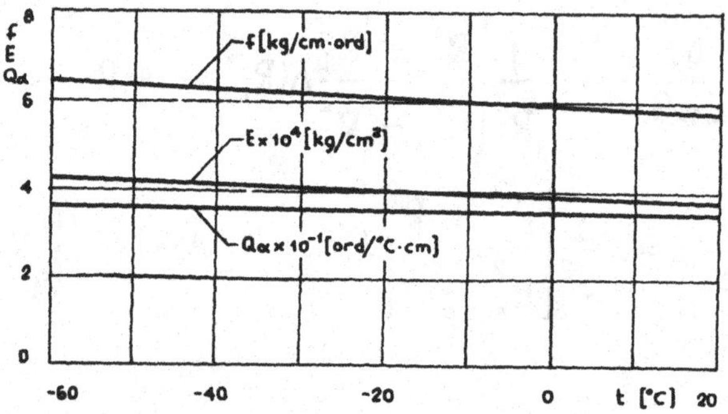

Fig. 4. Physical properties of Hysol 4290 as a function of temperature.

Epoxy resins are well suitable for photothermoelastic investigation owing to their high figure of merit Q_α .

Fig. 4 shows the Young's modulus E , the fringe value f and the figure Q_α of the epoxy resin Hysol 4290 as a function of temperature [14] .

Recording of Photoelastic Data

The photoelastic data are detected by using :

1) the standard polariscope;

2) the sandwich polariscope.

The standard polariscope is, generally, used for two-dimensional models.

It allows the determination of isochromatic fringes in models under stationary or transient temperature distribution and of isoclinic lines, but only under stationary conditions.

The sandwich polariscope is employed for three-dimensional models and for two-dimensional model in which it is necessary to prevent the axial flow of heat. It allows the determination of the isochromatic fringes and - at most - of only one isoclinic line (if a linear polariscope in dark field is used).

In order to have the whole field of isoclinic lines a set of tests with different orientation of the sandwich polariscope is necessary.

To avoid this, H. Becker and A. Colao in a recent paper [21] proposed a method - based on the isochromatic fringes - which allows the determination of the principal stresses direction in the sandwich polariscope.

The model (which must be symmetric in geometry and load) is made of two parts (separated from the axis of symmetry) one of which has a frozen stress in simple tension.

When the external load is applied, the frozen fringe order n_0 combines with external fringe order n_1 giving a resultant fringe order n which depends on the principal stresses direction. Measuring n_0, n_1 and n the direction of principal stresses is found.

Another method for the determination of the direction of the principal stresses is described by R. Mark and R. B. Pipes [22]. According to this method small holes are drilled in the model. The principal stresses direction is determined by the shape of fringes around the holes.

Separation of Principal Stresses

Generally two sets of data are available in two-dimensional photothermoelasticity, just like in the classic two-dimensional photoelasticity, that is :

1) the difference of principal stresses from the isochròmatic pattern;

2) the direction of principal stresses from the isoclinic pattern.

The principal stresses σ_1 and σ_2 can be separated :

1) using the equilibrium equations;

2) determining the sum of principal stresses (isopachic pattern).

Referring to the uncoupled quasistatic thermoelastic theory, the equilibrium equations are the same as these of isothermal elasticity. The methods of stress separation based on the Lamé-Maxwell equations of equilibrium (Filon method)

$$\frac{\partial \sigma_1}{\partial s_1} + \frac{\sigma_1 - \sigma_2}{\varrho_2} = 0$$

$$\frac{\partial \sigma_2}{\partial s_2} + \frac{\sigma_1 - \sigma_2}{\varrho_1} = 0$$

or on the carthesian equations of equilibrium (shear difference method)

$$\frac{\partial \sigma_x}{\partial x} + \frac{\partial \tau_{xy}}{\partial y} = 0$$

$$\frac{\partial \sigma_y}{\partial y} + \frac{\partial \tau_{xy}}{\partial x} = 0$$

can be therefore applied in photothermoelasticity as well as in two-dimensional photoelasticity.

The sum of principal stresses can be determined by means of the holographic method proposed by J. D. Hovanesian [23] or by using the compatibility equation [24]

$$\nabla^2(\sigma_1 + \sigma_2) = -E\alpha\nabla^2\theta \, . \qquad (4)$$

If the temperature distribution is harmonic equation (4) becomes

$$\nabla^2(\sigma_1 + \sigma_2) = 0 \, . \qquad (5)$$

That is, the sum of principal stresses too is harmonic, and can be easily determined by means of numerical methods or the electrical analogy.

If the temperature distribution in the model is not harmonic ($\nabla^2\theta \neq 0$) the equation (4) can be written

$$\nabla^2[(\sigma_1 + \sigma_2) + E\alpha\theta] = 0 \qquad (6)$$

In this case the sum of principal stresses is not harmonic, however the function $(\sigma_1 + \sigma_2) + E\alpha\theta$ is harmonic, consequently it can be determined by the isochromatic fringes and the temperature distribution along the boundaries using the numerical method or the electrical analogy.

The methods based on compatibility equation are useful especially if the isoclinic pattern is difficult to record. This happens (as already mentioned) in the study of transient thermal stresses and in the use of the sandwich polariscope technique.

Conclusion

Photothermoelasticity is a full field method which allows the study of thermal stresses by using the well-known technique of photoelasticity.

The accuracy of the results depends mainly on the availability of photoelastic materials with high figure of merit and having physical properties neraly constant in the range of temperature in which are tested.

References

[1] Slot, T. : "Photoelastic simulation of thermal stresses by mechanical prestraining", Exp. Mech., 1965 9 , 273-282;

[2] Khesin, G.I. - Strelchuk, N.A. - Shvey, E.M. - Savostyanov, V.N. : "Thermoelastic Stress Research by the Method of 'Unfreezing' Free Thermal Strains", Fourth Int. Conf. on Experimental Stress Analysis, 1970, Cambridge;

[3] Biot, M.A. : " A general property of two dimensional stress distribution", Philosophical Magazine 1935, Ser. 7 XIX, 540-549;

[4] Barriage, J.B. - Durelli, A.J. : "Application of a new deformeter to two dimensional thermal stress problems"; Proc. of S.E.S.A. 1956, XIII (2), 97-106;

[5] Ajovalasit, A. : "Determinazione fotoelastica, mediante l'analogia di Biot, delle tensioni termiche in cilindri con foro eccentrico soggetti a flusso di calore costante", Tecnica Italiana 1966, 12, 707-715;

[6] Gerard, G. - Gilbert, A.G. : "Photothermoelasticity : an exploratory study", Journal of Applied Mechanics 1957, 24(3), 355-360;

[7] Tramposch, H. - Gerard, G. : "Physical properties for photothermoelastic investigations", Journal of Applied Mechanics 1958, 25(4), 525- 528;

[8] Tramposch, H. - Gerard, G. : "An exploratory study of three dimensional photothermoelasticity", Journal of Applied Mechanics 1961, 28 (1), 35-40;

[9] Gerard, G. : "Progress in Photothermoelasticity", Proc. of the Internat. Symposium on Photoelasticity, Pergamon Press 1963, 81-93;

[10] Tramposch, H. - Gerard, G. : "Correlation of theoretical and photothermoelastic results on thermal stresses in idealized wing structures", Journal of Applied Mechanics 1960, 27 (1), 79-86;

[11] Belgrado, M. - Guerrini, B. - Vigni, P. : "Indagine fototermoelasticimetrica sullo stato di tensione in zone intagliate di piastre circolari con foro centrale ed altri fori periferici regolarmente intervallati soggette a data distribuzione di temperatura", Istituto di Impianti Nucleari, Facoltà di Ingegneria, Pisa, 1965;

[12] Gurtman, G. A. - Colao, A. A. : "Photothermoelastic investigation of stresses around a hole in a plate subjected to thermal shock", Proc. of

S.E.S.A. 1965, XXII (1), 97-104;

[13] Emery, A.F. - Barret, C.F. - Kobayashi, A.S.: "Temperature distributions and thermal stresses in a partially filled annulus", Exp. Mech. 1966, 12, 602-608;

[14] Ajovalasit, A. : "I dischi di turbina con palette saldate: tensioni termiche all'attacco delle palette", Tecnica Italiana 1968, 12 , 815-820;

[15] Ajovalasit, A. : "Photothermoelastic analysis of thermal stresses in discs with eccentric holes", Journal of Strain Analysis 1970, 5 (3), 223-229;

[16] Reichner, P. : "Stress concentration in the multiple-notched rim of a disk", Exp. Mech. 1961 11, 160-166;

[17] McQuillin, L.A. : "Experimental determination of thermal stress: a solid circular cylinder in transverse flow of hot gas", Journal of Strain Analysis 1968, 1, 1-10;

[18] Davis, J.B. - Swinson, W.F. : "Experimental Investigation of transient thermal stresses in a solid sphere", Proc. of S.E.S.A. 1968, XXV(2), 424-428;

[19] Burger, C.P. : "A generalized method for photo-elastic studies of transient thermal stresses", Proc. of S.E.S.A. 1969, XXVI(2), 529-337;

[20] Hovanesian, J.D. - Kowalski, H.C. : "Similarity in thermoelasticity", Proc. of S.E.S.A. 1967, XXIV(1), 82-84;

[21] Becker, H. - Colao, A. : "Determining Isoclinics from Fringe Patterns", Proc. of S.E.S.A. 1969, XXVI(2), 469-472;

[22] Mark, R. - Pipes, R.B. : "A simple method for determining principal stress directions in embedded-polariscope models", Exp. Mech. 1970, 9, 390-393;

[23] Hovanesian, J.D. : "New applications of holography to thermoelastic studies", Fourth Int. Conf. on Experimental Stress Analysis, 1970 Cambridge;

[24] Ajovalasit, A. : "La determinazione delle isopache nei casi di forza centrifuga e di tensioni termiche per mezzo di analogie elettriche", Tecnica e Ricostruzione (Boll. Ing. di Catania) 1970, 1.

INTERNATIONAL CENTRE FOR MECHANICAL SCIENCES

M. TSCHINKE
UNIVERSITY OF PALERMO

A SHORT SURVEY ON THE PROBLEMS CONNECTED WITH THE STUDY OF DYNAMICAL PHENOMENA BY MEANS OF PHOTOELASTICITY

UDINE 1970

COURSES AND LECTURES

F O R E W O R D

The Author is grateful to the Secretary General of CISM and to Professor V. Brčić for their invitation to give a lecture in the frame of the course on "Photoelasticity in Theory and in Practice".

The lecture is meant for reasearchers who have worked with Photoelasticity but have not used this tool to investigate dynamic phenomena.

In it the principal weaknesses of this particular branch of experimental stress analysis are examined, together with the ways by which some outstanding experts of the field have overcome them.

Some experimental results found by the Author are also given.

By dynamic photoelasticity we usually mean the study of the propagation of stresses with the aid of photoelastic techniques.

In order to do this one must have an ordinary transmission polariscope, a dynamic loading rig, which in the simplest case can be a falling weight, and some sort of high--speed photographic equipment.

The first dynamic research work has been done very soon after the discovery of photoelasticity as a tool for the design of structures and machine parts, yet today dynamic photoelasticity is still confined to a few research laboratories and must still be considered as being in a developmental stage.

The most important reasons for this situation are the following :

a) only plane stress fields can be examined by this means. The frozen-stress technique is obviously incompatible with dynamic loads. It is however possible to examine what happens in a given plane of a three-dimensional model, by using the "embedded polariscope" technique. Any discontinuity in the model can however cause reflection of waves, so this technique must be applied very carefully.

b) expensive and difficult-to-use equipment is usually necessary.

We will here consider only the types of cameras that are most commonly used, giving approximate figures for their prices, framing rates, number of frames obtainable.

Type of camera	framing rate (frames per second)	number of frames	price US $
Rotating prism	20.000	1.000	5.000
Stroboscope with drum camera	50.000	500	10.000
Multiple spark (Cranz-Schardin)	1.000.000	40	20.000
Streak	c o n t i n u o u s		15.000
Single flashes with increasing delays	2.000.000	no limit	7.500
Rotating mirror (Beckman & Whitley)	2.000.000	50 - 100	25.000

This kind of equipment is difficult to use because the cameras and especially the synchronizing equipment con - necting the camera to the loading rig and to the light source, are rather delicate and complicated.

We will not try to describe the loading rigs, as they are usually built especially for the job.

c) <u>Transferability of results from model to prototype.</u> This is by far the greatest difficulty. It derives from the fact that for all materials Hooke's law is not applicable under rapidly changing load conditions. Structural materials can be, generally speaking, considered as having a non-linear stress-strain equation, depending on the rate of load variation, which means that they have hysteresis, whereas model materials have a time-dependent, linear differential equation, which means that they can be considered, in general, as being linearly visco elastic. Moreover the experimental constants (photoleastic constant) are also dependent on the strain rate.

This gives an idea of the difficulties involved. The use of models is however essential in dynamic photoelasticity, because even the best cameras cannot easily follow the propagation of stresses in structural materials, owing to their high moduli. For this reason the photoelastic-coating technique can not usually be applied. With this technique there are also other difficulties due to reflections of waves, as pointed out by Khesin et al.

We must conclude therefore that dynamic photoelasticity can be safely used only to examine stress propagation in photoelastic materials.

One way out is given by Kuske [1] , who has developed a similarity criterion based on the internal damping of the material. He measures the "half-value-length" (Halbwertslänge) of the material, i.e. the distance a stress wave of the same shape as that foreseen for the prototype has to travel in the material in order to reduce its amplitude by half. If this parameter is the same for model and prototype, the model must be built in a 1 : 1 scale and the results will be transferable. The above value is in the order of 1 to 3 m for model material and of about 1, 000 m for steel. This means that model sizes will be extremely small; especially when machine parts are studied. In these cases the only thing to do is to build models in a convenient size and to examine only the first part of the dynamic phenomenon, during which the internal damping has no great influence.

Another possible type of work is the determination of dynamic stress-concentration factors, as done by Manzella [2] which seem to be transferable within certain limits to prototypes. It is well known that stress concentration factors usually tend to decrease with increasing strain rates.

d) Separation of principal stresses. Kuske [1] uses the following generalized equilibrium equations :

$$\frac{\partial \sigma_x}{\partial x} + \frac{\partial \tau_{xy}}{\partial y} + \varrho \frac{\partial^2 s_x}{\partial t^2} = 0$$

$$\frac{\partial \sigma_y}{\partial y} + \frac{\partial \tau_{xy}}{\partial x} + \varrho \frac{\partial^2 s_y}{\partial t^2} = 0$$

and gives the complete procedure for the separation.

The knowledge of the isoclinics field is, of course, required in this case, and this is a further complication.

Kuske also proposes a method based on the compatibility condition.

Experimental methods like the oblique incidence method used by Flynn and others and holographic interferometry methods are certainly more efficient, since they permit to record simultaneously all the data needed for the separation of the principal stresses.

Classification of Impacts

According to Kuske [3] one can have three types of impacts, depending on the value of the ratio of the duration of contact t_0 (t_0 depends chiefly on the ratio m_1 / m_2 of the mass of the impacting body to that of the impacted body) to the period of vibration of the impacted body, t_1

quasistatic for $t_0/t_1 > 1$

vibrational for $t_0/t_1 \cong 1$

dynamic for $t_0/t_1 \ll 1$

in the last case we have $t_0 \cong t_2$ time needed for the stress wave to cross the model (t_2 depends chiefly on the modulus of the impacted body and on its dimensions).

Many researchers have concentrated on the last type of impacts, since only in purely dynamical conditions it is possible for the stress waves to influence strongly the stress field.

Keeping in mind the above definitions one has to conclude that purely dynamic impacts are likely to occur mostly in large structures or in the crust of the earth and not, for instance, in machine parts.

Elements of the three types can however be present simultaneously or in subsequent times in most impact phenomena.

Experimental Results

We will now give some results that have been obtained in Palermo [4] using Hysol 4485 and impacts that were between the second and the third type in the above classification.

It is known that the longitudinal or dilatational wave travels in a plane plate with a velocity given by

$$C_L = \sqrt{\frac{E}{\varrho(1-\mu^2)}}$$

Meyer, Nagamati and Taylor [5] have shown that C_L is, photoelastically speaking, the velocity of the zero-order fringe and that higher-order fringes must have lower velocities. The fringe-propagation velocities for the 0, 5, 1, 5 and 2, 5 order fringes have been measured for various impact velocities, keeping the mass-ratio constant. The velocity of the zero-order fringe was obtained by extrapolation. The results are given in Fig. 1, together with the values of the dynamic modulus, calculated by means of the expression of C_L. The theoretical value of C_L based on the static value of E is 60,11 m/s.

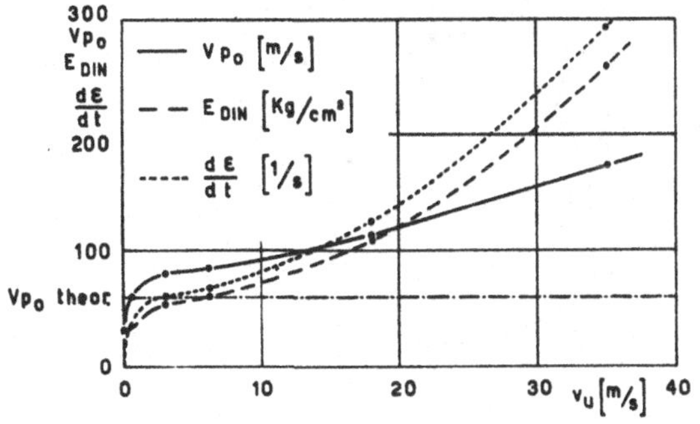

Fig. 1. Propagation velocity of the zero-order fringe (V_{po}), dynamic modulus (E_{din}) and strain rate ($d\epsilon/_{dt}$) in a plate of Hysol 4485, all plotted against the impact velocity.

The values of the strain-rate $\frac{d\varepsilon}{dt}$ also plotted in Fig. 1, were arrived at indirectly through a law of dependency of E on $\frac{d\varepsilon}{dt}$ given by Dally, Riley and Durelli [6].

The dependency of the shape of the isochromatics on the impact velocity has also been investigated.

According to Durelli and Riley [7] the stress waves during this first phase of the impact follow the pattern given in Fig. 2, which shows the P-wave (dilatational) and the S-wave (distortional) as well as the two secondary distortional waves S_1' and S_2' generated by the P-wave as it travels along the boundary. This means that the 0,5 order fringe (the first visible one in our case) will reproduce the shape of the P-wave except near the boundary, where it will be deformed by the presence of the S' waves. Higher-order fringes will be the result of the superposition of the three types of waves.

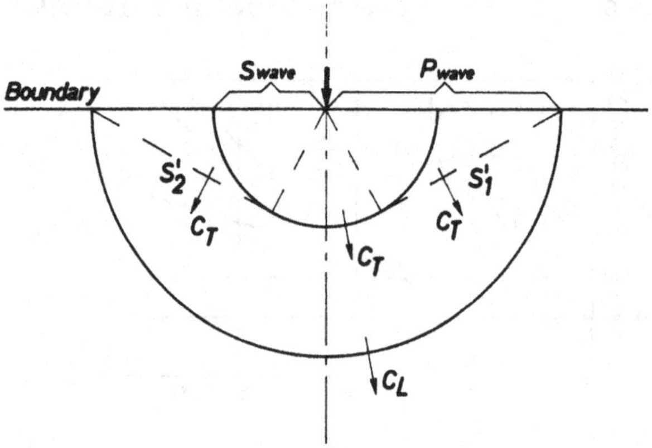

Fig. 2: Position of the stress waves during impact according to Durelli and Riley [7].

We have found that the shape of the isochromatics depends on the impact velocity. Fig. 3 shows four fringe patterns photographed at approximately the same time after impact, with four different impact velocities.

Slow impacts produce a pattern still very similar to the Hertzian one, with nearly circular, "pear-shaped" fringes. Two small, separate 0, 5 order fringes on the boundary contain two isotropic points. With higher impact velocities these points are "overtaken" by the main 0, 5 order fringe, which becomes bell-shaped. With still higher velocities the isotropic points fall within the 1, 5 and the 2, 5 order fringes. This derives from the fact that, while the fringe velocity increases

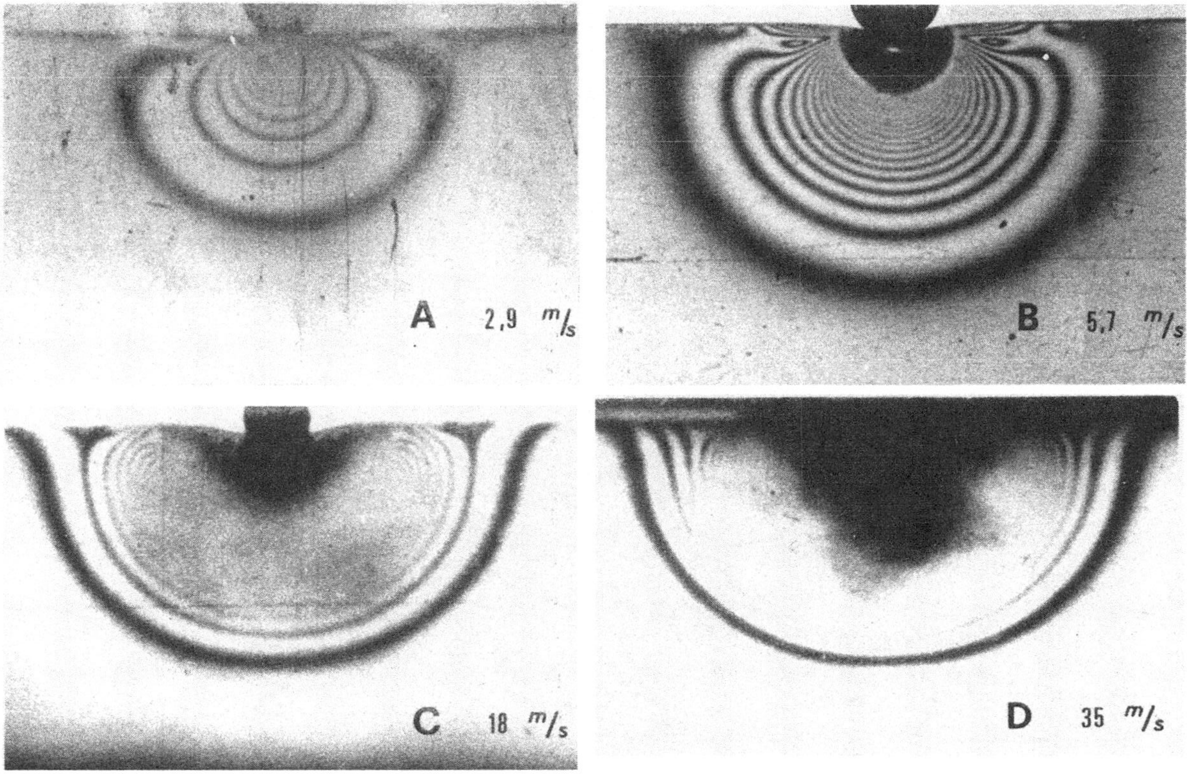

Fig.3. Fringe patterns during impact for four different impact velocities. Elapsed time after the beginning of the impact approx. 0,5 ms for all four images.

with the impact velocity, the rate at which the isotropic points move remains practically constant for all impact velocities.

Fig. 4 shows the fringe shapes, expressed by the ratio of the axes, plotted against the impact velocities.

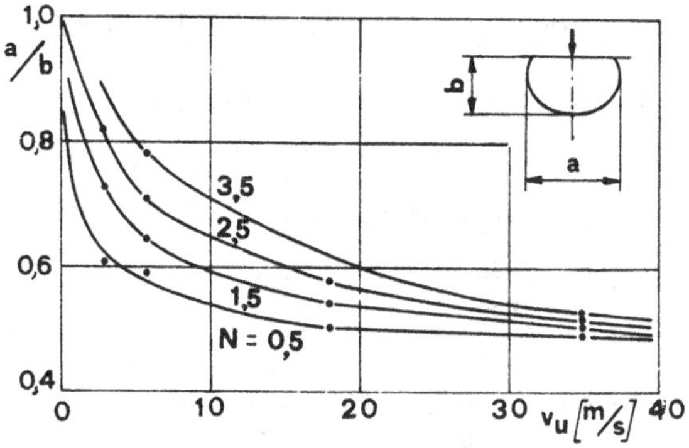

Fig. 4. Dependence of the shape of the fringes, expressed by the ratio a/b of their axes, on the impact velocity.

It must be kept in mind that for each velocity we have a different impact energy, due to the fact that the m_1 / m_2 ratio was kept constant. We are planning to do equal-energy and equal m_1 / m_2 tests, which means having to work with different models for each impact velocity.

Another feature that has been examined is the duration of contact t_0 for a short, simply supported beam. It was found that this value decreases slowly as the impact velocity increases, which is in accordance with Hertz's theory of contact. The decrease is approximately proportional to $v_{imp}^{-\frac{1}{9}}$ whereas according to Hertz it should be proportional to $v_{imp}^{-\frac{1}{5}}$. This difference is certainly due to the fact that Hertz was considering a perfectly elastic material.

Betser and Frocht [8] working with a higher-modulus material, Castolite, have found that in similar conditions the duration of impact does not change for different impact velocities. This is in accordance with the elementary theory of De St. Venant and Flamant for long beams.

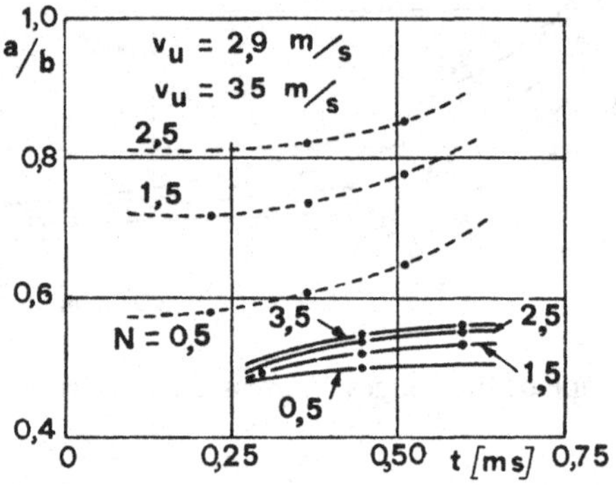

Fig. 5. Values of the ratio a/b Vs. time for two different impact velocities.

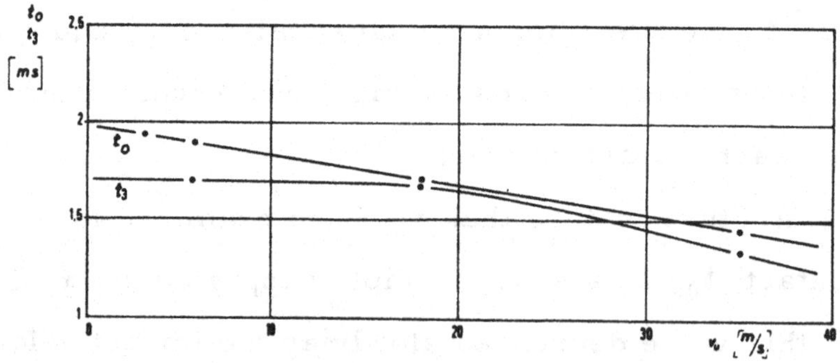

Fig. 6. Duration of impact (t_0) and time lapse (t_3) between the beginning of the impact and the occurence of the maximum fringe order, shown Vs. the impact velocity.

References

[1] Kuske, A. : "Contributo allo studio della propagazione di stati di tensione con l'ausilio del procedimento fotoelastico"
Paper given at the "Seminario sul tema "Analisi sperimentale delle tensioni nei recipienti a pressione", Bologna, Faculty of Engineering, September 15-18, 1970;

[2] Manzella, G. : "Fattori di concentrazione delle tensioni e coefficienti di forma per aste forate soggette ad urto trasversale". Tecnica Italiana, Anno XXIX, No. 4, Aprile 1964;

[3] Kuske, A. : "Photoelastic research on dynamic stresses"
SESA, Vol. XXIII, No. 1, p. 105;

[4] Meyer, M. L.; Nagamati, G.; Taylor, D. A. W. : "Notes on photoelastic observations of impact phenomena", Post-graduate Dept. of Applied Mechanics, Sheffields University;

[5] Tschinke, M. : "Urto trasversale su travi corte"; Disegno di Macchine, Palermo, No. 1, 1969;

[6] Dally, J. W.; Riley, W. F.; Durelli, A. J. : "A photoelastic approach to transient stresses problems employing low-modulus materials". ASME Paper No. 59-A-10;

[7] Durelli, A. J.; Riley, W. F. : "Introduction to photo-mechanics"
Prentice-Hall Inc. Englewood Cliffs, N. J. 1965;

[8] Betser, A.A.; Frocht, M.M. : "A photoelastic study of maximum tensile stresses in simply supported short beams under central transverse impact".
ASME Paper No. 57-APM-36.